THE CUSTOMER-DRIVEN COMPANY:

Managerial Perspectives on Quality Function Deployment

Second Edition

William E. Eureka

Nancy E. Ryan

Dearborn, Michigan

IRWIN

Professional Publishing

Burr Ridge, Illinois
New York, New York

Printed in the United States of America

Senior sponsoring editor: Jean Marie Geracie
Production manager: Laurie Kersch
Designer: Heidi J. Baughman
Compositor: Montgomery Media, Inc.
Typeface: 12/15 Bookman
Printer: Book Press, Inc.

Library of Congress Cataloging-in-Publication Data

Eureka, William E.
The customer-driven company: managerial perspective on quality function deployment/William E. Eureka, Nancy E. Ryan.
p. cm.
Includes bibliographical references and index.
ISBN 0-7863-0141-4
1. Product management. 2. Quality function deployment. 3. Quality of products. 4. Quality function deployment—United States. I. Ryan, Nancy E., 1959- . II. Title.
TS156.E86 1994
658.5'62—dc20 93-38371

1 2 3 4 5 6 7 8 9 0 10 9 8 7 6 5 4

Contents

Preface

Quality Function Deployment (QFD) represents the synthesis of numerous methodologies developed in America but perfected and integrated by the Japanese. It is not steeped in theory—rather, it was developed by users and is a continually evolving methodology that can be adapted to meet a variety of industrial- and business-related needs.

QFD helps us focus on the details of our business—whether it is manufacturing- or service-oriented—which results in success in the marketplace. It helps us focus our energies on the high-risk details that often go unfulfilled and allows our standard operating system to manage the myriad of lower-risk details. Most importantly, QFD helps us identify and meet customer needs and produce quality- and cost-driven products in an era when not doing so results in slipping market share.

Implementation of QFD in America began in the automotive industry but is now rapidly spreading to all major sectors of industry. I have personally witnessed the successful implementation of QFD in both domestic automotive and nonautomotive environments and am looking forward to many more success stories as the benefits of QFD continue to be realized. And I am confident that the creative nature of American managers and engineers will result in innovative applications of QFD that far exceed what the Japanese have accomplished.

Although the first edition of *The Customer-Driven Company: Managerial Perspectives on QFD* was published in1988, the preceding paragraphs from it are just as relevant today as they were six years ago. Since that time, a number of the predictions that were made in that initial preface have come to pass.

Implementation of QFD in America has spread from the automotive industry to all major sectors of U.S. industry, and QFD applications—both in manufacturing- and service-oriented environments—have resulted in an invigorating sense of renewal and resourcefulness and an essential restructuring of "how we get things done." The QFD phenomenon was just beginning in 1988 and, as expected, QFD has since become a mainstay of American industry. It is not a flash in the pan.

Today, there are many success stories that depict the benefits to be achieved with QFD, several of which are featured in this revised edition. This updated and expanded edition also contains current material that reflects six solid years of growth and learning with QFD, as well as a new chapter entitled "Customizing Your House." This chapter features advanced aspects of QFD, five sample QFD applications, and eight basic steps to QFD implementation.

A glossary of QFD-related terms and a comprehensive bibliography of QFD resources have also been added. In addition, the flow of the book has been altered to reflect these changes and in response to the voice of the customer; that is, readers who suggested improvements to the first edition. The appendix on Taguchi Methods—a highly touted complement to QFD trademarked by the American Supplier Institute—remains.

As noted in the first edition's preface, QFD is especially valuable in that it brings out the best in a number of other tools and methods, including Total Quality Management (or Companywide Quality Control);

Statistical Quality Control (or Statistical Process Control); Taguchi Methods and robust design; Fault Tree Analysis and Reverse Fault Tree Analysis; Failure Mode and Effect Analysis; the Kano model; Pugh Concept Selection; and simultaneous or concurrent engineering.

American managers and engineers are exploiting this asset of QFD in some very innovative ways, as the case studies at each annual Symposium on Quality Function Deployment attest. And because of QFD's "what-if" nature, we will undoubtedly see even more novel uses of QFD in the coming years.

In conclusion, I'd like to again cite from the preface to the first edition: This book—an easy-to-read overview of QFD for today's busy manager—will not tell you everything there is to know about QFD. What it will tell you is what America's QFD pioneers think and have learned about the methodology; what QFD basically is, involves, and does; and how, where, and when QFD can best be utilized. Hands-on training and application must follow.

William E. Eureka

Acknowledgments

The authors would like to thank the following people for their contributions to this book:

Michael E. Chupa, vice president of marketing at ITT Automotive, Auburn Hills, Michigan;

Dr. Don Clausing, Bernard Gordon Adjunct Professor of Engineering Innovation and Practice at Massachusetts Institute of Technology, Cambridge, Massachusetts;

Earl C. Conway, retired corporate director of quality worldwide at Procter & Gamble, and Adjunct Professor of Industrial Engineering, University of Cincinnati, Cincinnati, Ohio;

Robert J. Dika, quality planning executive—small car platform, customer satisfaction and vehicle quality office at Chrysler Corporation, Auburn Hills, Michigan;

Akashi Fukuhara, director of Central Japan Quality Control Association, Nagoya, Japan;

James T. Gipprich, sales manager of the light duty vehicle brake group at Kelsey–Hayes, Romulus, Michigan;

Calvin W. Gray, president of the Handy & Harmon Automotive Group, Auburn Hills, Michigan;

Walton M. Hancock, William Clay Ford Professor of Product Manufacturing at the University of Michigan, Ann Arbor, Michigan;

Kevyn Irving, manager of technical services and contract development at Ethicon Endo–Surgery, Cincinnati, Ohio;

Ralph A. Kirkpatrick, program manager of CS6-8OC2 total product quality at GE Aircraft Engines, Cincinnati, Ohio;

Norman E. Morrell, corporate manager of quality-product reliability at The Budd Company, Troy, Michigan;

George R. Perry, vice president and general manager at Siemens Automotive, Newport News, Virginia;

Robert M. Schaefer, director of systems engineering methods at General Motors, Warren, Michigan;

Raymond P. Smock, retired manager of advanced quality concepts development and product assurance at Ford North American Automotive Operations, Dearborn, Michigan;

Peter J. Soltis, retired senior technical specialist of product engineering at Kelsey–Hayes, Romulus, Michigan.

Chapter One

Sizing Up the Competition

In the late 1980s, when the first edition of *The Customer-Driven Company: Managerial Perspectives on QFD* was published, trade friction and the threat of protectionist measures led to an international invasion of North American soil. Direct foreign investments gave birth to new manufacturer-supplier relationships, new factories, and new American jobs. Richard Florida (1992), professor of management and public policy at Carnegie Mellon University, states that "foreign direct manufacturing investment has played at important role in the region's [the Heartland's] economic revival." He adds that a large concentration of the world's best U.S., Japanese, and European companies now call the region home.

An article titled "Foreign Investment Spurs Region's Employment Prospects," published in *The Christian Science Monitor,* 24 March 1993, reported the following: In Marysville, Ohio, where a transplant Honda plant is located, the per capita income more than doubled between 1980 and 1990. The average 1990 income was $30,547, the highest in Ohio's 1988 counties, and housing starts increased nearly 20-fold.

But while Marysville and its surrounding areas prospered, some critics claim that Detroit bore the brunt of the foreign invasion. Others claim that such companies as

Honda of America are American companies, and others claim that the 100 percent American company is as extinct as the dinosaur. Conversely, subsidiaries of major American corporations are manufacturing their products on foreign soil—using parts exported from the United States—and dipping into pockets of foreign market share, further clouding the balance-of-trade issue. Should such products be called imports or exports? That is the question, with the answer contingent on who is doing the asking—and the answering.

Welcome to the global 90s, where traditional distinctions between domestic and foreign competition are blurry, blurry, and blurred. As the multinational company comes of age, operating independently and in joint ventures on several continents, one might surmise that American industry could loosen its tie. This, however, is not the case. The thorny issues of the trade deficit, rumored dumping violations, and protectionist antics continue to draw blood. Furthermore, American market share is no longer a measure of success, with almost every American industry participating in the ongoing war of the global marketplace.

While trade friction between the United States and both Japan and the European Community continues to escalate, and industry and government attempt to negotiate fair practices in the new global arena, the companies that are surviving—even thriving—are those that most quickly respond to the ever-changing voice of the customer—that is, customer-driven companies.

A WHOLE NEW BALL GAME

The global 90s present both challenges and opportunities—to perfect management and manufacturing practices, to satisfy customer and company requirements simultaneously, to manufacture products at lower cost with higher quality, and to manufacture such products sooner than the competition.

Quality Function Deployment (QFD), the focus of this book and an essential ingredient of the customer-driven company, can help turn these opportunities into realities by reducing the time to bring a product to market by one-third to one-half (and thus reduce total product costs and result in a more timely, competitive product). It will also reduce material costs and overhead, the two highest contributors to total manufacturing costs. QFD does all this by minimizing engineering changes within the product development cycle and by ensuring that required changes occur on paper rather than on the factory floor. According to Walton M. Hancock (1988), William Clay Ford Professor of Product Manufacturing at the University of Michigan in Ann Arbor,

> We have to really focus our ability to produce high-quality products at low cost and to produce products people want to buy in a world market. The combination of the two together is very crucial. The ability to produce things people want to buy in a world market means that products have to be designed so that people worldwide will want them—not just people in the U.S. The other issue is the ability to change products quickly—if your competitor comes out with something better, you'd better be able to react to it.

Financial gurus Austin H. Kiplinger and Knight A. Kiplinger (1989) claim that American industry is on the rebound, but it still faces several challenges. Citing a study made by the Council on Competitiveness, a private organization, Kiplinger and Kiplinger note that

> Many nations, instead of trying to match America's scientific prowess, focus on acquiring new technology, rapidly translating it into commercial translations, then making improvements in response to market signals.

> The phenomenal success of that strategy, pioneered by Japan and followed by other countries, makes it clear that U.S. industry cannot expect to prosper by playing a design role, inventing state-of-the-art products that can be duplicated elsewhere in a matter of months at a lower cost.

Rather, U.S. industry needs to beat these foreign countries at their own game. Consider, for example, a new thermoplastic deployment door developed for the 1992 Lincoln Town Car passenger air bag by a team led by John Boag and Chuen-Chu (Teresa) Lang of Ford Motor Company. The air-bag deployment door, which is expected to save Ford $1.6 million a year, won Boag and Lang the 1992 Henry Ford Technology Award and took first place in the Society of Plastics Engineers Most Innovative Use of Plastics Award. It evolved from a team effort between Ford and DuPont, which supplied the two-component thermoplastic material, DYM 100.

According to an article in the March 1993 issue of *Ford World*,

> By using the new door, Ford more than halved injection molding times, shaved the door's weight by half a pound, and trimmed $12 from the unit's cost. Put another way, if 138,000 vehicles use the new system in one year, Ford saves more than $1.6 million.

In addition, and perhaps even more importantly, the new door meets a customer requirement for safety devices on both the driver and passenger sides, a customer requirement that had stumped suppliers and was expected to be cost prohibitive. Such a product embodies the spirit of QFD.

The QFD spirit is also reflected in such customer-driven products as the 1993 Ford F-150 Lightning pickup and the Mustang Cobra, the first two high-performance vehicles developed by Ford's special vehicle team. These new vehicles are designed to meet driving enthusiasts' specific customer requirements. By bypassing time-consuming traditional channels, the special vehicle team refined both vehicles, which were introduced at about $2,500 more than their base model counterparts, in two years of intense teamwork.

An era of industrial downsizing and restructuring, the global 90s, however, are as much about services as they are about products, especially since the majority of new jobs created in the 1990s will be in the service industries. The service sector, too, will be competing globally, with many kinds of services being highly exportable (Kiplinger and Kiplinger 1989).

A special report in *Fortune* magazine on U.S. productivity notes the following regarding competition for services.

> In the Nineties, the knives are out. Big chunks of the service economy—telecommunications, transportation, banking, insurance—have seen government regulation lifted or loosened, destroying cushy markets. Often the competition is global. For example, foreign banks' share of commercial and industrial loans to U.S.-based business jumped from 18 percent in 1983 to 45 percent by the end of 1991.

QFD shares respect for the customer with leading service providers and recognizes that today's customers demand better services for their money. QFD is as applicable to the service arena as it is to product-development and continuous-batch processes.

INDUSTRIES UNDER ATTACK

When Kiplinger and Kiplinger published *America in the Global '90s* in late 1989, they noted that although the United States still holds the overall lead in science and technology, the gap is closing, with other nations drawing close and even overtaking the United States in a few specific fields. Technologically speaking, the European Community now appears much better equipped to compete and could very well emerge as a high-tech power by the year 2000. But the main competition still comes from Japan. "Together, the U.S. and Japan will account for half of global high-tech sales," Kiplinger and Kiplinger reported, "with the U.S. maintaining its lead."

The Kiplinger team cited a U.S. technological lead in the following industries: aerospace, supercomputers, microprocessors, computer software, medical technology, food technology, and bioengineering. The Japanese, on the other hand, are leading the race in factory automation, robotics, ceramics, semiconductors, high-definition TV, and other areas of consumer electronics. Superconductivity, fiber optics, composites, and telecommunications are too close to call. The integrated European Community also accounts for changes in the global marketplace, particularly the telecommunications, apparel, financial services, and electronics industries, according to the Kiplinger team.

Made in America: Regaining the Productive Edge is based on a study conducted in the late 1980s by the MIT Commission on Industrial Productivity at the Massachusetts Institute of Technology. It examined the status of eight key American industries: automotive; chemical; commercial aircraft; consumer electronics; machine tools; semiconductors, computers, and copiers (microelectronics); steel; and textiles. The commission found that five of the eight industries (automotive, consumer electronics, machine tools, steel, and textiles) have been seriously wounded by foreign competition, with the chemical, commercial aircraft, and microelectronics industries facing increased competition. Considering the long-term ramifications of the above projections, the need for a time-saving, cost-reducing, and quality-enhancing tool like QFD has never been greater.

THE GLOBAL PERSPECTIVE

Changing Alliances (1987) was published as a result of the Harvard Business School Project on the Auto Industry and the American Economy. The authors, Dyer, Salter, and Webber, reported that globalization poses a direct challenge to both the American automotive industry and the American economy, because the American system has a fundamental disadvantage in the free-world market. Other industries will experience similar plights as well, the study noted, as globalization sweeps American industries.

Current trends in the automotive and aviation industries support the Harvard Business School study. While the Big Three automakers continue to address the issues raised by globalization, the aviation industry is caught in a double bind, with decreased Department of Defense spending and increased competition from Airbus Industrie, a consortium of European companies, which has met the customer need for increased width of airplane aisles. Such leading U.S. companies as Boeing, McDonnell-Douglas, and United Technologies are responding to the aviation industry's burgeoning recession and market changes by slashing their workforces and making drastic production cutbacks.

These factors, not to mention projected production costs, are behind a potential global venture between Boeing and four members of rival Airbus. They are conducting a feasibility study of a jumbo jet that could carry from 550 to 800 passengers and fly from 8,050 to 11,500 miles nonstop. In the case of this jumbo jet, the old adage "If you can't beat them, join them," makes perfect sense. The world is not ready—or big enough—for more than one jumbo jet.

The jumbo jet, with a turn-of-the-century start-up date, would replace the Boeing 747 as the world's largest airplane and cost $10 billion to $15 billion to produce. Such high costs have led both Boeing and Airbus to approach other international players regarding their participation in the project. Including such players as Mitsubishi Heavy Industries and Kawasaki Heavy Industries would make the proposed collaboration truly global in nature.

The U.S. aluminum industry has also been adversely affected by going global. It has shut down 10 of its 32 aluminum smelters in the last 15 years and has sold technology to foreign competitors that benefit from lower energy and labor costs.

According to the article titled "U.S. Aluminum Makers Hit Downside of Going Global," published in *The Wall Street Journal*, 22 March 1993,

> While American companies in many industries, from chemicals to consumer products, have been affected recently by troubles overseas, aluminum makers have a bigger problem because they have more international exposure. . . . Since 1978, aluminum prices have been set on the London Metal Exchange, putting U.S. producers at the whim of overseas traders.

Although cost-cutting and marketing efforts are expected to come to the ailing industry's rescue, such efforts were less than successful when applied to the U.S. steel industry.

The computer industry, another hotly contested global market, is also experiencing future shock in the global 90s. In their book *Computer Wars: How the West Can Win in a Post-IBM World,* industry analysts Charles Ferguson and Charles Morris examine how "Big Blue" (IBM) won and lost global technological dominance as the personal computer overtook the mainframe. Noting that the IBM of the late 1980s began courting profits instead of change—and virtually ignored fundamental changes in the global computer market—Ferguson and Morris contend that the global computer war between the United States and Japan will be won by the Silicon Valley model of business management; that is, companies that thrive and survive on the cutting edge from which the profits flow. The smaller, fast-footed computer companies now challenging IBM must "push themselves to develop new computer architectures every few years so that the Japanese can't get a single product architecture completely in their sights before another is developed," according to Mark Clayton in *The Christian Science Monitor,* 30 March 1993.

IBM once embraced change as a means of staying atop of the global computer market and responding to customer needs. When it slowed its response to change and innovation—keys to America's industrial success since the days of the Model T—its competitiveness suffered.

From industry to industry—automotive, aviation, aluminum, and computer, to name just a few—the ability to fend off the competition and develop new technology is critical to survival and growth in today's global marketplace. And these are just two areas in which QFD shines.

VISIONS FOR THE FUTURE

The Mondeo—a small sedan with a big price tag, taking five years and $6 billion to produce—has been called the first truly global product, that is, a world car. With the Mondeo, Ford aggressively addressed the controversy of designing a single product to suit all the world's markets. The automobile was first introduced in Europe and then, with slightly different styling, in North America. Even the Mondeo's name has a global ring to it. It is based on the word for *world* in the Romance languages.

The article "Big Bet: Ford Is Turning Heads With $6 Billion Cost To Design 'World Car,'" in the *The Wall Street Journal*, 23 March 1993, attests that the Mondeo embodies the need for a new global strategy—and for a customer-driven process like QFD.

> In the past, Ford's huge, autonomous European and North American auto operations designed and built completely different models, reflecting sharply divergent needs in the two markets. But now tastes are converging, Ford contends, and the company says it can no longer afford to duplicate efforts.

Ford is reportedly encouraged by early sales of the Mondeo, which *The Detroit News*, 18 April 1993, described as "a high-margin car that promises to pad Ford's bottom line and give it a chance to reestablish supremacy in the midsized car market." Ford's European market share improved three percentage points upon the car's introduction in Europe.

The *ISO 9000 Compendium: International Standards for Quality Management* notes that the newly united European Community places new marketplace pressures on companies that wish to trade or even compete with European companies in other markets. Although quality systems and technology for hardware products are quite mature, "the most rapid development of new quality technology will occur in the other three general product categories [software, processed materials, and services] during the 1990s."

In addition, the ISO's vision for the future involves an across-the-board migration toward product offerings that combine two or more generic product categories; that is, products with processed materials that are incorporated into manufactured parts or hardware assemblies in which computer software also is an incorporated feature. Additionally, the service aspects of selling, delivering, and servicing are important features of the total package.

As this chapter clearly illustrates, changes in global marketplace trends abound. Yet one thing is certain. Today's corporate world—an increasingly complex, combative, and competitive place—is moving too fast for companies with Rip Van Winkle tendencies. Companies that increased efficiency and reduced costs in the 1980s must continue to do so, making continuous improvement, which QFD can certainly help achieve, their objective.

REFERENCES

Byland, K. 1993. "New Air Bag Wins Henry Ford Tech Award." *Ford World.* (March): 7.

Byland, K. 1993. "Special Vehicle Team Launches Cobra, Lightning." *Ford World.* (March): 6.

Dertouzos, M.L., R.K. Lester, R.M. Solow, and The MIT Commission on Industrial Productivity. 1989. *Made in America: Regaining the Productive Edge.* Cambridge, Mass.: MIT Press.

Dyer, D., M.S. Salter, and A.M. Webber. 1987. *Changing Alliances.* Boston: Harvard Business School Press.

Ferguson, C., and C. Morris. 1993. *Computer Wars: How the West Can Win in a Post-IBM World.* New York: Times Books.

Florida, R. 1992. *Rebuilding America: Lessons from the Industrial Heartland.* Pittsburgh: Carnegie Mellon University.

Hancock, W.M. 1988. Telephone interview. Interviewed by Nancy Ryan.

International Organization for Standardization. *ISO 9000 Compendium: International Standards for Quality Management.* Geneva, Switzerland: International Organization for Standardization.

Kiplinger, A.H., and K.A. Kipplinger. 1989. *America in the Global* '90s. Washington, D.C.: Kiplinger Washington Editors.

Stewart, T.A. 1993. "U.S. Productivity: First but Fading." *Fortune,* 19 October, 54–7.

Chapter Two

Turning Things Around

Documentation of how the Japanese turned their manufacturing operations around has been zealously sought since the mid-1970s. With this pursuit came a greater understanding of Japan's industrial rejuvenation and the operating differences between Japanese and U.S. companies (see Figure 2–1). As American engineers, managers, and quality control professionals made study missions to Japan, they saw that the Japanese did more than copy western managerial and technological concepts and techniques. The Japanese religiously studied these tools and selectively applied the ones that seemed to work best—and were best suited to their specific cultural environment.

The Japanese didn't set out to become inventors. Rather, their approach smacked of common sense—with a bit of innovation added for good measure. The managerial- and manufacturing-related procedures debated were not new to the manufacturing realm. When deemed appropriate to the situation, they were applied. When not deemed appropriate, they were discarded. Many could just as easily have been applied in the United States or another country. Application of other procedures would have required a bit more forethought. In some cases, cultural aspirations and expectations might have made them less than effective.

The use of Statistical Quality Control (SQC) and Statistical Process Control (SPC) had many merits; the

Operating Differences

Japan	U.S.
Deploy the voice of the customer (decide what is important)	Specify internal requirements (everything is important)
Design and build to target values (reduce variation or dispersion)	Design and build to specification tolerances (manage around tolerance stack)
Optimize product and process design	React to customer problems

FIGURE 2–1
A number of operating differences exist between Japanese and U.S. companies; in particular, identifying and deploying customer requirements (QFD) and optimizing product and process design (Taguchi Methods).

use of extensive work-in-process inventories did not. The teachings of Dr. W. Edwards Deming and Dr. Joseph M. Juran were put into practice; Just-in-Time Manufacturing would be applied later. Teamwork, consensus, and organization were desirable; irrationally complex systems were not. Manufacturers began working in concert with unions, suppliers, and governments; simplicity became a word to work by.

The results of such efforts included improved scheduling and production control systems, process technologies, quality systems and procedures, internal communications, and problem-solving techniques. By fashioning a management and engineering philosophy that produced quality products with reduced variation through optimized design and improved processes, many Japanese companies accumulated market share. A long-term effort that resulted in continuous quality improvement, and what began with statistical quality methods, evolved into Total Quality Control (TQC), which itself matured to become Companywide Quality Control (CWQC) and, later, Groupwide Quality Control (GWQC).

WHAT'S RIGHT, WHAT'S WRONG

According to Robert J. Dika (1993), quality planning executive—small car platform of the customer satisfaction and vehicle quality office at the Chrysler Technology Center,

> The Japanese weren't successful because they were culturally different, had better workers, or even had economic advantages. They were successful because they were systematic and behaved sensibly as an organization. They operated as a team focused on necessary results and the keys to success.

And just as the Japanese incorporated western concepts and techniques, American companies are now applying the same selective-reasoning process to such Japanese tools as QFD, tailoring them to American needs and improving them wherever possible.

Chrysler, for example, began using QFD in 1986. Seven years later, at the Fifth Symposium on Quality Function Deployment, Dika (1993) shared the following anecdote.

> On a bright day in February, a colleague of mine was walking down the hall and was stopped by the general manager of small car platform, with whom he had not exchanged more than a few words in some time. This

general manager had supported his people and QFD work teams in the face of persistent rumors that QFD was being discouraged by top management. He was beaming and anxious to talk, knowing that my friend was a QFD advocate. He had just returned from a weekend customer research event in San Diego, where the PL, now the new Dodge and Plymouth Neon, still a year from introduction, had been evaluated.

Not only did the customers like the Neon, but they were overflowing with praise. They said that they could imagine a young married couple on vacation, in the PL with the windows down, having fun driving up the California coast. In their explanation of what they liked about the car, the research participants kept mentioning terms like *well made, value, fun to drive,* and *roomy*—the very attributes that three years ago our QFD project had identified as critical.

Following the research, the cars were shown to a group of writers at the Chrysler Arizona Proving Grounds just outside of Phoenix. The executive engineer of vehicle development made a presentation on the QFD study, he talked about how we identified up front what we wanted to do and how we kept measuring our progress against the customer's requirements, particularly in the area of fun to drive. The press people were similarly excited, both about the Neon and the changes that we had made in our development process.

The general manager then, with a look of delight, took my friend into his confidence. "You know, this really works!" he said.

When the customers' requirements aren't met, however, the results can be devastating. The circumstances surrounding the May 17, 1987, Iraqi jet attack on the USS *Stark*, which killed 37 American sailors, attest to this fact. When a radar operator became annoyed by the beep of an electronic warning system designed to alert the *Stark* crew of hostile radar, he turned the system's audible signal off. The system's visual signal wasn't detected in time, and the *Stark* tragedy was partially blamed on human error. That human error, however, resulted from an unhappy customer.

"The tale is a nightmare for the manufacturers of sophisticated electronic weaponry, who find increasingly that the systems they are building have become too complex for soldiers and sailors to operate properly," wrote John H. Cushman, Jr., in *The New York Times,* 21 June 1987. "Moreover, maintenance and repair of electronic weapons are also often beyond the ability of most military personnel, despite the investment of enormous amounts of training time, according to experts both inside and outside the government."

IN RETROSPECT

Industry gurus have earmarked a number of inefficiencies that result in unmet customer needs. Planning, although usually performed with the best intentions, is often performed hastily. This occurs because planning is performed under the restraints of a conflicting message: It's important to plan but nonproductive to do so. Consequently, planning provides general direction but little attention to detail, which is left to the execution stage. There's no time to do things right—but plenty of time to do things over.

Objectives receive similar handling. Objectives are generally construed as worthwhile goals, but the means to achieve them are often unclear. Differences in interpretations and priorities, hidden conflicts, and so forth all add up to objectives that aren't clearly defined.

Product development and manufacturing inefficiencies don't help either. The former include losing sight of the customer, preoccupation with the schedule instead of the product, inadequate front loading, improper use of product testing, new but not better designs, preoccupation with design changes, isolation of design efforts, designing and building to specification tolerances instead of target values, inadequate consideration of manufacturing needs, and achievement of sensitive optimums. Manufacturing inefficiencies include excessive inventories, added cost to improve quality, and dependence on operator-sensitive procedures. Focusing on short-term problem solving and using high technology for problem solving result in addition roadblocks. Figure 2–2 contrasts this problem-solving approach with the more lucrative Japanese approach of problem prevention.

According to an insightful *Fortune* magazine cover story,

> There should be no mystery to managing smarter. A European commission that examined U.S. productivity in the early 1950s marveled at how the Yanks did it—and their observations sound a lot like what Americans say today about the Japanese. "American operatives do not work harder than their opposite numbers in Britain, nor are their machines in general better tooled or superior," said a British team. "Their high productive efficiency is due largely to more accurate planning by management and more constant analysis of methods."

The MIT Commission on Industrial Productivity (1989) contends that industrial America must focus on continually improving its product design and production processes and encouraging company-to-customer, company-to-supplier, and company-to-company cooperation, as well as cooperation between industry and government.

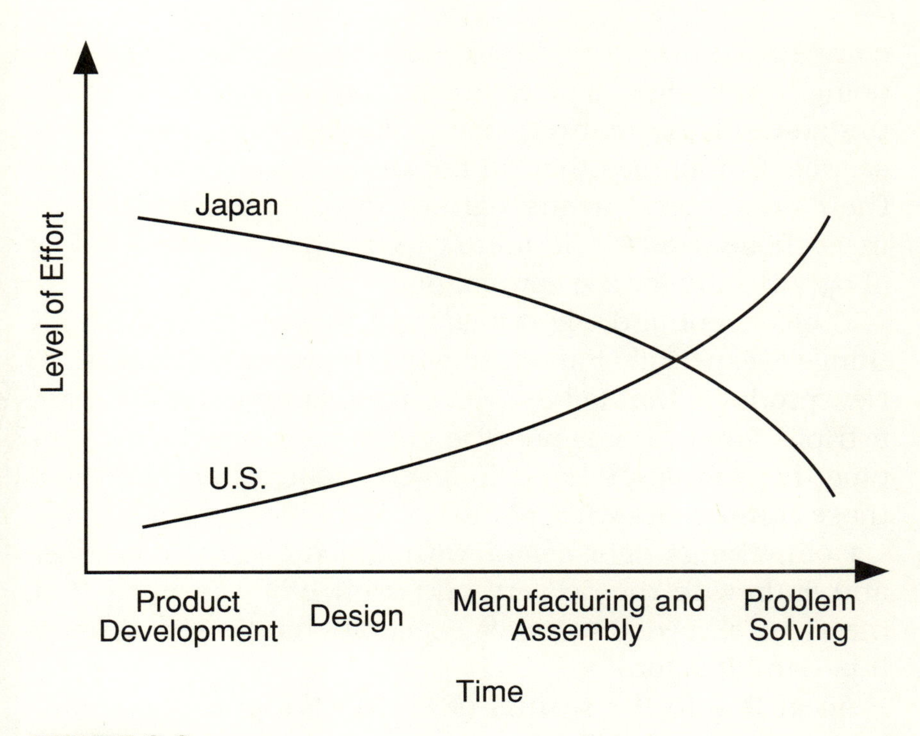

FIGURE 2–2

During the product development process, Japanese engineers focus their efforts on such upstream activities as research and development and design. U.S. engineers, on the other hand, typically concentrate on problem-solving activities.

Now the good news: More American companies are doing the right things—reorganizing their production processes, allotting more responsibility to employees, instituting training in such tools as Total Quality Management, QFD, and SQC, and creating new liaisons with suppliers and customers.

In today's competitive global marketplace, however, it's not enough to just do the right things. In order to become competitive and remain so, companies must focus on doing things right. For example, companies must totally understand the marketplace needs for their product or service. Companies that do not understand and embrace their customers' needs cannot develop products that meet those needs. And failure to do so creates a window of opportunity for the competition.

The concept of doing things right versus doing the right things is especially important when it comes to innovation. New products that fail to meet customer expectations pose a triple threat, damaging the company's reputation, tipping the company's hand to its competitors, and showing those competitors what not to do.

Doing things right begins with a focus on the customer and ends with the delivery and service of a product that meets or exceeds customer expectations. Sound simple? It is—and it's not.

According to the results of Project Sappho, a comprehensive industry research project conducted in Great Britain in the early 1970s, the reason that some products succeed and others fail begins with an "an imaginative understanding of customer needs done well early." This leads to successful products and shortened development times.

Used in this context, imaginative can be defined as an in-depth understanding of the customer's needs for the product, how the product is going to be used, and the noise factors that must be considered in order to make the product robust or insensitive to factors that cannot be controlled.

In contrast, unstable product definitions, that is, product definitions not based on customer expectations, typically result in failed products with extended product development times. The latter usually results from constantly redefining the product requirements as new information, which should have been determined at the onset, is uncovered.

THE 10 CASH DRAINS

Dr. Don Clausing (1988), the Bernard M. Gordon Adjunct Professor of Engineering Innovation and Practice at MIT, has identified 10 cash drains that plague U.S. industry. These cash drains, which companies should obviously avoid if they wish to make money, are the following.

Technology push—but where's the pull? The United States is very good at technology generation, but a threefold problem exists: (1) new technological concepts are developed and major resources spent, yet no discernable customer need can be identified; (2) strong customer needs exist for which technology-generation activities are lacking; and (3) good technological concepts are developed for which there are clear customer needs, but these concepts are inadequately transferred into system design activities.

Disregard for the voice of the customer. Products are often doomed to mediocrity at the first step in system design—determination of customer needs. Contributing to this failure is emphasis on the voice of the executive or the voice of the engineer, rather than the voice of the customer.

The "great idea." Product concept selection often results after someone shouts, "I have this great idea!" This concept then becomes the only concept given serious consideration, although it may be highly vulnerable and unable to withstand the test of time.

Pretend designs. Pretend designs are new but not better product designs. Often, they're not production-intent designs, focusing instead on the creation of experimental prototypes. A lack of production intent leads to a disastrous attitude: "This is just a first design, I'll fix it later."

Pampered products. Product concepts are often pampered so that they look good in demonstration. This approach has been improved upon by rigorous application of reliability growth and problem-solving methodologies. These, however, aren't adequate approaches to optimization of vital design parameters.

Hardware swamps. Hardware swamps occur when prototypes are so numerous, with overlapping test phases, that excessive time is spent debugging and maintaining them rather than improving the product design.

Here's the product, where's the factory? If production capability design starts only a few months before actual production then many severe problems will occur.

"We've always made it this way." Process-operating points (for example, speeds, depth of cut, feed rates, pressures, temperatures, and so forth) are specified on process sheets or numerical control programs. Process parameter values often have been fixed for a long time and may have resulted from little, if any, development. This leads to a dangerous attitude: "We've always made it this way and it works." This, in turn, leads to reliance on tradition rather than innovation.

Inspection. Factory inspection, sorting out the good products from the bad after production is complete, is an outdated, inefficient process. The same holds for product testing (design inspection) during development.

"Give me my targets, and let me do my thing!" Early allocation of targets to a detailed level tends to destroy teamwork. Contracts where each person works in isolation lead to subsystems that can't be integrated, products that can't be produced, production capacity that can't produce modern products, operating systems that attempt to enslave their users, managers who can't seem to manage, and employees who wait to be told what to do.

According to Clausing, an improved total development process, which includes QFD, aimed at continuously bringing better products to market will plug the cash drains. "One difficulty is getting people to admit that there are problems in the work process," he added. "But the thought is slowly permeating U.S. industry that our work process needs improving."

THE ROLE OF QFD

When Japanese companies deploy the voice of the customer, they mobilize all employees to focus on continuous quality improvement with reduced costs and faster response times. In the broadest sense, CWQC refers to quality of management, human behavior, work performance, work environment, products, and service, which combined, encompass quality of society, industries, the national economy, and global competition.

For maximum long-term benefit, QFD should be applied as the main product development thrust of CWQC. QFD shouldn't be used on every part of every product, or even on every product. To do so would defeat one of QFD's main thrusts—bringing a high-quality product to market in as short a time span as practical. Instead, priority is given to the parts and functions with the most potential for improved competitive advantage.

QFD should be used to identify and focus on the high-risk details of product development. The normal operating system at the typical American company can successfully handle the majority of product development details. QFD should only be applied to the product aspects that the normal system cannot ensure: problem areas and the implementation of innovations.

In Japan, QFD is often applied in conjunction with a specific companywide theme, such as reducing product development time, reducing product defects, and so forth. According to Akashi Fukuhara (1988), director of the Central Japan Quality Control Association, most Japanese winners of the Deming Prize use QFD. The Deming Prize is awarded to the competing company that achieves the most distinctive performance improvement through the application of SQC in a given year.

American companies, too, are now tailoring QFD to their needs and benefiting from the customer-driven procedure. For instance, Florida Power & Light Company, which won the Deming Prize in 1990, uses QFD. (See sidebar on page 89.)

A myriad of other American companies are also using QFD. Some are product-oriented organizations and others are service based. According to Michael E. Chupa (1993), vice president of marketing at ITT Automotive,

> As we see it, during this phase of supply base contraction, where there are too many companies chasing too few opportunities, QFD is a more and more important

part of our strategic toolbox. We're using it to pinpoint our approach to upcoming platforms vis-à-vis customer expectations and competitor offerings.

REFERENCES

Center for the Study of Industrial Innovation. 1972. "Project Sappho (Scientific Activity Predictor from Patterns of Heuristic Origins)." In *Success and Failure in Industrial Innovation.* London: University of Sussex.

Chupa, M.E. 1993. Letter to Nancy Ryan.

Clausing, D.P. 1988. Personal interview. Interviewed by Nancy Ryan.

Dertouzos, M.L., R.K. Lester, R.M. Solow, and The MIT Commission on Industrial Productivity. 1989. *Made in America: Regaining the Productive Edge.* Cambridge, Mass.: MIT Press.

Dika, R.J. 1993. *QFD Implementation at Chrysler: The First Seven Years.* Paper read at the Fifth Symposium on Quality Function Deployment, 21–22 June, at Novi, Michigan.

Fukuhara, A. 1988. Personal interview. Interviewed by Nancy Ryan.

Richman, L.S. "How America Can Triumph." *Fortune,* 18 December, 52–4+.

Chapter Three

How QFD Came to Be

It's time to update the old widget story. Imagine this: A company has just introduced a new widget—at half the cost and twice the productivity and quality of the competition's widgets. Contributing to this feat was QFD—a system for translating customer requirements into appropriate technical requirements at each stage of the product development process—and the engineering tools it specifies.

Half the cost and twice the productivity and quality in two-thirds the time; that's QFD in action, as the following real-life example illustrates. It is based on firsthand information gleaned from a study mission to Japan. Toyota Motor Corporation's primary transmission supplier, Aisin A.W. (formerly Aisin Warner), a subsidiary of Aisin Seiki in Kariya, Japan, used QFD to reduce the number of engineering changes during product development by half. Development time and start-up cycles were also cut in half by enhancing the overall time to market. Numerous other companies are now using QFD with similar results. They are—or soon will be—the competition.

What exactly is QFD? Taken literally, the term *Quality Function Deployment* may seem a bit misleading. QFD isn't a quality tool—although it can certainly improve quality in the strictest sense of the word. Rather, it is a visually powerful planning tool. And although first used by the Japanese, several aspects of QFD resemble Value

Analysis/Value Engineering (VAVE), a process developed in America, combined with simple marketing techniques.

Broadly speaking, QFD is a system for translating customer requirements into appropriate company requirements at each stage of the product development cycle. QFD is used from research and development to engineering, manufacturing, marketing, sales, and distribution (see Figure 3–1).

The term is derived from six Chinese and Japanese characters: *hin shitsu* (qualities, features, or attributes), *ki no* (function), and *ten kai* (deployment, development, or diffusion), as Figure 3–2 illustrates. The translation is inexact, as well as nondescriptive of the actual QFD process, as *hin shitsu* is synonymous with qualities (that is, features or attributes) instead of quality.

QFD AND QUALITY DEFINED

George R. Perry (1988), vice president and general manager of Siemens Automotive in Newport News, Virginia, defined QFD as "a systematic way of ensuring that the development of product features, characteristics, and specifications, as well as the selection and development of process equipment, methods, and controls, are driven by the demands of the customer or marketplace."

Many other QFD practitioners have drafted their own definitions of QFD as well. These include the following:

- A sophisticated Pareto analysis.
- A fundamental, commonsense approach to product development that focuses on proactive rather than reactive quality control.
- A technique to help neutralize the voice of the customer.

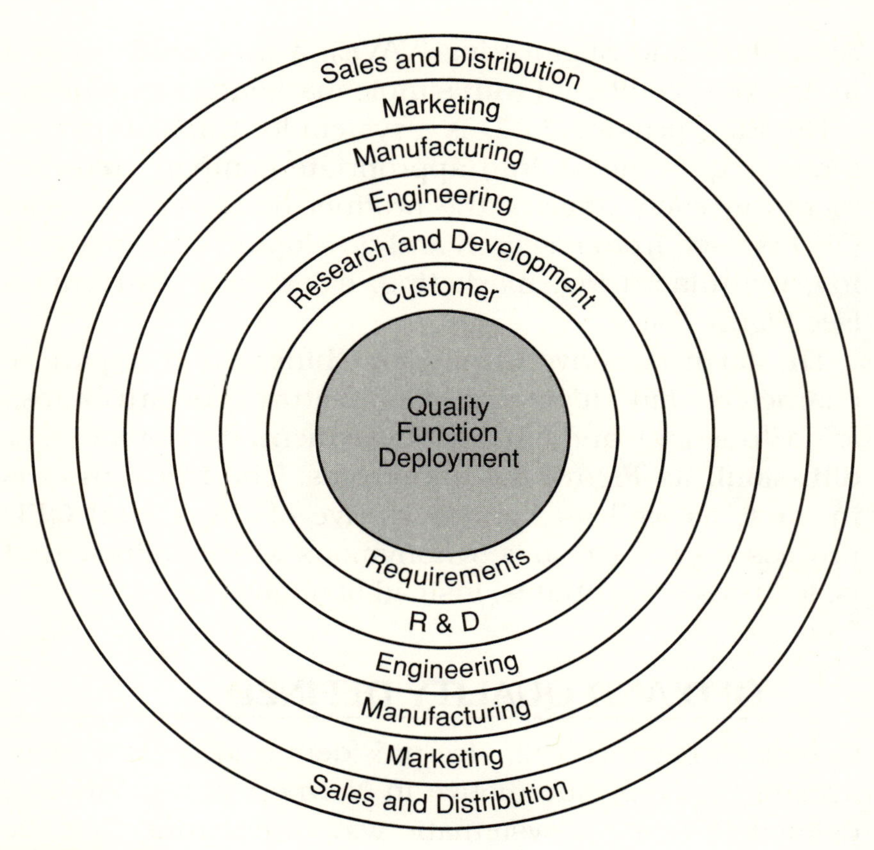

FIGURE 3–1

QFD translates customer requirements into appropriate company requirements at each stage of the product development process.

- A systematic way of documenting and breaking down customer needs into manageable, actionable details.
- A planning methodology that organizes relevant information to facilitate better decision making.
- A framework for customer-derived product development objectives.

FIGURE 3–2
The term *Quality Function Deployment* is derived from six Chinese and Japanese characters: *hin shitsu* (qualities, features, or attributes), *ki no* (function), and *ten kai* (deployment, development, or diffusion). The translation is inexact: *Hin shitsu* is synonymous with qualities, not quality.

- A way of reducing the uncertainty involved in product and process design.
- A technique that promotes cross-functional teamwork.
- A methodology that gets the right people together early on to work efficiently and effectively to meet customer needs.

A number of definitions describe the term *quality* as well, not to mention the term *quality control.* One of the foremost goals of the majority of today's companies is to produce high-quality products or to provide high-quality service. But what, precisely, is a high-quality product or service? Ask the question more than once and you'll undoubtedly get conflicting answers. And do high-quality products always turn a profit? Unfortunately not. More elementary yet, what is quality in action; that is, what is quality control? To some, it's conformance to specification limits. To others, it's much more.

The *ISO 9000 Compendium: International Standards for Quality Management* provides the following definition of quality:

> The totality of features and characteristics of a product or service that bear on its ability to satisfy stated or

implied needs. [Quality control] concerns the operational means to fulfill the quality requirements, while quality assurance aims at providing confidence in this fulfillment, both within the organization and externally to customers and authorities.

According to *Merriam Webster's Collegiate® Dictionary* (10th edition), quality control is "an aggregate of activities (as design analysis and statistical sampling with inspection for defects) designed to ensure adequate quality, especially in manufactured products."

Another standard reference source, the *McGraw-Hill Dictionary of Scientific and Technical Terms,* prefers this definition of quality control. "Inspection, analysis, and action applied to a portion of the product in a manufacturing operation to estimate overall quality of the product and determine what, if any, changes must be made to achieve or maintain the required level of quality."

The *Glossary and Tables for Statistical Quality Control,* published by the American Society for Quality Control (ASQC) defines quality control as "the overall system of activities whose purpose is to provide a quality of product or service that meets the needs of users; also, the use of such a system."

Note that key word, *users,* in the ASQC definition. With users in mind, quality becomes more comprehensive than the characteristics usually associated with it, encompassing performance, extra features (added options), reliability, durability, serviceability, aesthetics, and conformance to standards. It even transcends time when associated with such issues as safety and environmental protection.

Like beauty, quality is in the eye of the beholder—and the beholder should be the customer. Hence any definition of quality should be supplied by the customer, which is precisely what QFD ensures. QFD communicates the customer's perception of quality via the voice of the customer.

FOCUS ON PROBLEM PREVENTION

Industrial leaders using QFD deploy the voice of the customer to help determine important product attributes. Engineers at these companies then design and build to target values, seeking to reduce manufacturing variation around these targets. They focus on optimizing the product and the process, not only to maximize performance, but also to reduce variation. This results in consistently high performance, from product to product and throughout the product's lifetime.

By front loading their product development efforts, these companies focus on planning and problem prevention, not problem solving. QFD is one of the methodologies used to make the transition from reactive to preventive, from downstream, manufacturing-oriented quality control to upstream, product-design-oriented QC (see Figure 3–3). It does so by defining what to do and how to do it in a manner that results in consistent performance that satisfies customers.

By clearly defining the job objectives needed to achieve customer-defined quality, QFD helps build quality into a product. And while it doesn't guarantee success, QFD greatly improves the probability of achieving it. Without QFD, you get what you've always gotten. With QFD, you get a new, improved approach to product planning. You also get protection from the unknown unknowns—those little surprises that visit the product development process when you least expect them. By raising a multitude of questions at a project's onset, QFD rapidly turns those

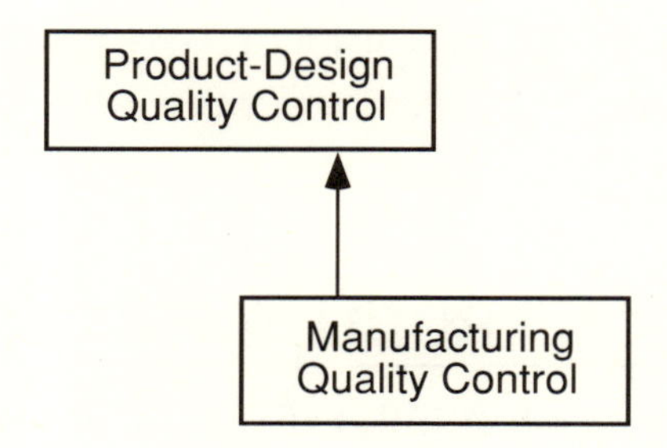

FIGURE 3–3
QFD results in an upstream, product-design-oriented—versus a downstream, manufacturing-oriented—form of quality control.

unknown unknowns into known unknowns, which you can then look out for (see Figure 3–4).

IMPROVED PRODUCT DEVELOPMENT

QFD brings out the best in a variety of engineering tools already available. These include competitive benchmarking, Failure Mode and Effect Analysis (FMEA), Fault Tree Analysis (FTA) and Reverse Fault Tree Analysis (RFTA), Taguchi Methods and robust design, Pugh Concept Selection, and VAVE. When properly applied, these tools, which are defined in the glossary, will help ensure quality products.

With QFD, broad product development objectives are broken down into specific, actionable assignments using a comprehensive team effort. Without this team approach, QFD loses much of its power. The process is accomplished via a series of matrices and charts that deploy customer requirements and related technical requirements from product planning and product design to process planning and the shop floor (see Figure 3–5).

On a short-term basis, QFD results in fewer start-up problems, fewer design changes, and shorter product

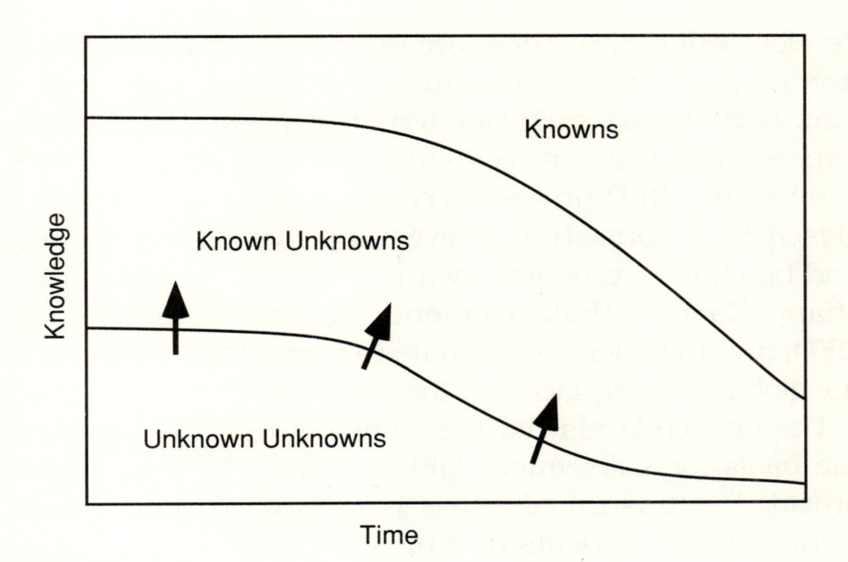

FIGURE 3–4
QFD raises a multitude of questions at the onset of a project, thus turning unknown unknowns into known unknowns.

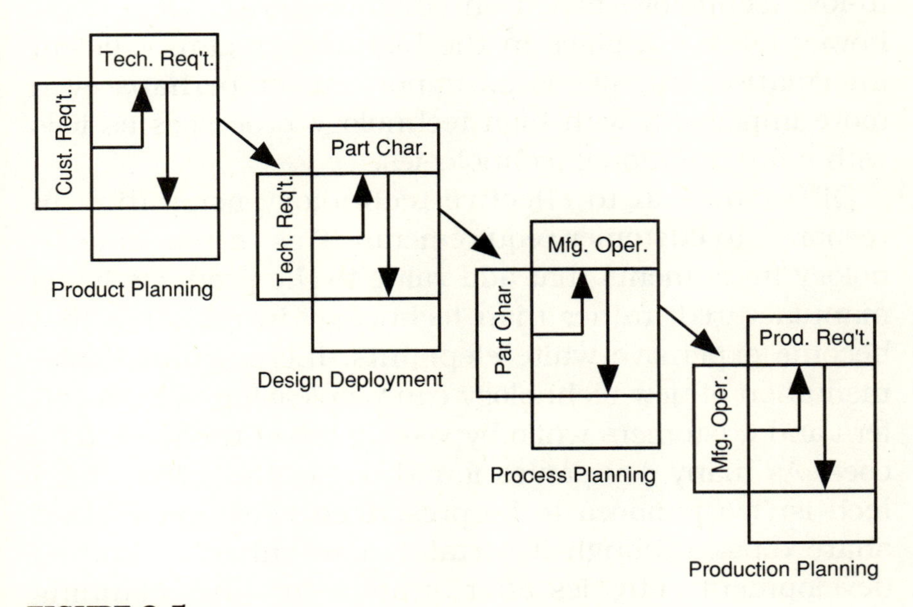

FIGURE 3–5
QFD matrices and charts deploy customer requirements and related technical requirements all the way down to the shop floor.

development cycles. These are essential for improved engineering productivity and reduced costs. Even more important, however, are such long-term benefits as satisfied customers, lower warranty costs, and increased market share.

When the QFD process is correctly utilized, it creates a closed loop consisting of ever-improving cost, quality, and timeliness; productivity and profitability; and market share. Each of these elements figures prominently in QFD, and together, they equate to competitive strength in the global marketplace (see Figure 3–6).

The first QFD matrix, the House of Quality, serves as the basis for subsequent QFD phases. The information provided in this initial phase is used to identify specific technical requirements that must be achieved in order to satisfy customer requirements. (The mechanics of the total QFD process will be detailed in chapter 4).

QFD is not a high technology; rather, it is a medium-to-low technology based on common sense. QFD does, however, have a place in the high-tech realm: Efficient information transfer is as important, or perhaps even more important, with high technology processes as it is with more traditional technologies.

QFD can lead to effective technology generation in response to customer requirements. This results in technology investments that add value to the products being manufactured, rather than technology investments that become expensive white elephants. Incremental implementation of new technology can be pleasing to both coffers and customers while bypassing toy-of-the-week fiascoes. As many companies found out in the 1980s, high tech isn't a panacea to be prescribed whenever market share dips, although it certainly can enhance product development activities after strategic product planning has occurred.

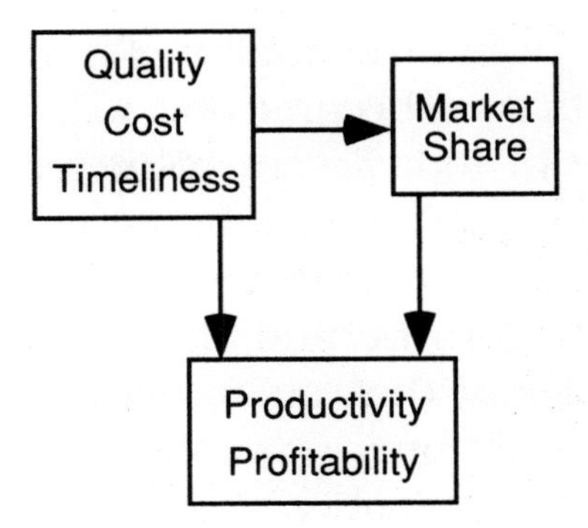

FIGURE 3–6
The use of QFD results in improved cost, quality, and timeliness, which result in increased productivity and ultimately, market share.

"We have to become good at what I call medium tech," explained Don Clausing (1988) of MIT. "We tend to want to retrench to high technology because we feel it's where we have an edge." Clausing, previously principal engineer in the advanced development activities of Xerox, was introduced to QFD in March 1984 while visiting Fuji–Xerox in Tokyo. His research into the company's product development process led to a meeting with one of its primary consultants, Dr. Hajime Makabe. After returning from Japan, Clausing shared his knowledge of QFD with engineers from Ford. Next came a series of annual American Supplier Institute (ASI) study missions to Japan that focused increased attention on QFD.

The methodology, which has been used in America with increased regularity and success since the mid-1980s, was formalized at Kobe Shipyard, Mitsubishi Heavy Industries, in Kobe, Japan. Although the shipyard built one massive, sophisticated ship at a time, the benefits of a strategic planning system that detailed and documented the relationship between the quality of a finished product and its components soon became obvious.

This holds true in today's era of increased customer sophistication, increased international competition, and increased technological innovation. These three elements are the primary factors behind the escalating product variety and complexity that customers now demand in diverse areas ranging from automobiles to telecommunications to diapers. Companies that use QFD can more efficiently respond to these quick-changing customer requirements with the latest technological developments, and they can do so in record time.

REFERENCES

American Society for Quality Control Statistics Committee. 1983. *Glossary and Tables for Statistical Control.* Milwaukee: ASQC Quality Press.

Clausing, D.P. 1988. Personal interview. Interviewed by Nancy Ryan.

International Organization for Standardization. ISO 9000 *Compendium: International Standards for Quality Management.* Geneva, Switzerland: International Organization for Standardization.

Lapedes, D.N., Ed. 1978. *McGraw-Hill Dictionary of Scientific and Technical Terms.* New York: McGraw-Hill Book Co.

Perry, G.R. 1988. Personal interview. Interviewed by Nancy Ryan.

Chapter Four

The QFD Approach

"Back when a knight went to a specialized blacksmith to have a coat of armor constructed, things were much simpler—the customer was speaking directly to the blacksmith, who could then do Quality Function Deployment in his head," explains MIT's Don Clausing (1988).

In today's complex industrial environment, however, the customer and the shop floor operator manufacturing the product seldom speak to each other. At a small, family-operated shop, the voice of the customer is heard because dealing with the customer is everyone's concern. But as the family-operated shop grows, walls crop up between specialized departments and the customer. QFD helps punch holes in these walls so that the voice of the customer can come through. In other words, QFD serves as an intermediary, bringing the voice of the customer directly to the shop floor.

The voice of the customer is characterized by customer requirements defined by detailed consultation, brainstorming, feedback mechanisms, and market research. QFD deploys the voice of the customer through the total product development process. This involves translating customer requirements into appropriate technical requirements for each stage of product development and production (see Figure 4–1).

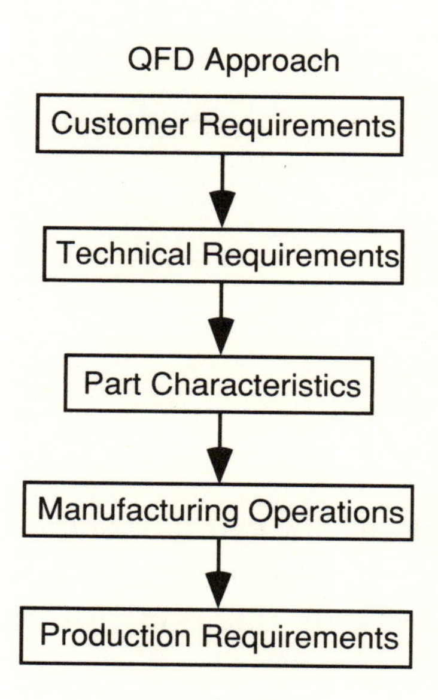

FIGURE 4–1
QFD translates customer requirements into appropriate technical requirements, part characteristics, manufacturing operations, and production requirements.

Developing a top-notch project team is one of the most challenging—and potentially rewarding—aspects of the process. All areas involved in product development should be represented on this team: marketing, product planning, product design and engineering, prototyping and testing, process development, manufacturing, assembly, sales, and service. And all should be working toward a shared goal—a customer-defined product to be completed by a specific date and at a specific cost. If there's any magic in the QFD process, it's the magic of teamwork: Even in cases where QFD has been less than successfully applied, the resulting teamwork is typically considered as successful.

LAYING THE FOUNDATION

The basic approach used in QFD is conceptually similar to the practice followed by most manufacturing companies. The process begins with customer requirements, which are usually loosely stated qualitative characteristics, such as looks good, is easy to use, works well, feels good, is safe, is comfortable, lasts long, is luxurious, and so forth. These characteristics are important to the customer but often defy quantification and are difficult to act on.

Verbalized customer requirements are often called *performance quality.* Not all customer requirements are verbalized, however, especially those that relate to hidden components and assemblies. In such cases, the customer expects some basic product functions, also known as *basic quality.* In addition to basic quality, today's customer has come to expect *excitement quality,* or pleasant surprises that he or she has never experienced before but nonetheless wants or needs. The Kano model plots the type of quality (that is, performance, basic, or excitement) on a grid representing customer satisfaction and degree of achievement (see Figure 4–2). This model can be used to help differentiate between these very different customer requirements and to obtain a creative understanding of customer needs.

In other words, the Kano model points toward a product that not only satisfies the customer, but delights him or her. In fact, 3M calls these pleasant surprises customer delights.

During product development, customer requirements are converted into internal company requirements called *company measures.* These measures are generally global product characteristics that will satisfy customer requirements if properly executed. Product development does

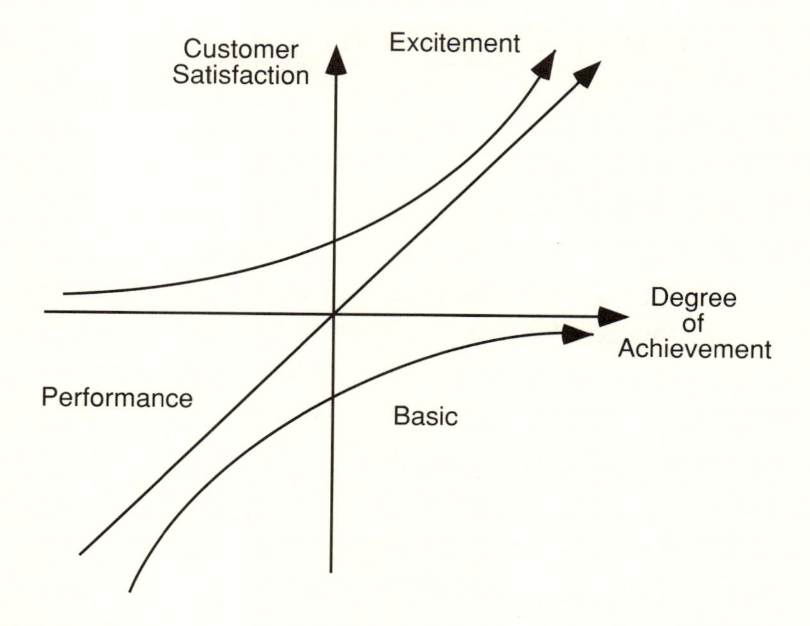

FIGURE 4–2

The Kano model plots the type of quality (that is, performance, basic, or excitement) on a grid representing customer satisfaction and degree of achievement. It can be used to differentiate between customer requirements and to obtain an imaginative understanding of customer needs.

not usually begin at this global level, however; rather, it begins at the system, subsystem, or parts level. Global company measures are hence translated into critical *part characteristics* that allow the essential functions of the product to be performed.

The use of the word *part(s)* here and in the following sections is appropriate for products that are assemblies of mechanical components. QFD applies equally well to other types of products, including almost any combination of ingredients, materials, or services. *Ingredients, materials, services,* or other relevant terminology may be substituted for *part(s)* in this and subsequent discussions.

Determining the required manufacturing operations is the next step, one that is often constrained by previous capital investments of factories and equipment. Within these operating constraints, the *manufacturing operations* most critical to creating the desired part characteristics are determined, as are the process parameters of the most influential operations.

The manufacturing operations are then translated into the *production requirements* shop floor personnel will use to consistently produce the required part characteristics. These include inspection and Statistical Process Control (SPC) plans, preventive maintenance programs, operator instructions and training, and mistake-proofing devices for preventing inadvertent operator errors. This is an entire set of procedures and practices that will aid in the manufacture of products that will ultimately satisfy customer requirements.

This hierarchical approach is not unlike the approach many companies have used for years with varying degrees of success. But problems occur when some of the translations are not properly made. There are several key reasons for these improper translations, including large organizational structures and complex product development processes.

American companies are normally structured with strong vertical line organizations and fairly clear reporting hierarchies. When a new program of great importance is implemented, the lines of many departments must be spanned, forming the horizontal linkages necessary to complete the program. The vertical linkages, however, are often so strong that departmental loyalties are at odds with the program requirements.

According to the *Fortune* review of the MIT study *Made in America* (1989),

> American firms tend to have hierarchies arranged as an organizational tree; people working in different departments often have to go up the tree to their lowest-level common superior and back down. Information flows slowly, if at all, from marketing to R&D to production. Professionals have difficulty working in teams with specialists in other disciplines. Decisions that should be integrated instead are made sequentially. . . . In Japanese firms, by contrast, the hierarchy has fewer levels. People in one layer generally know people in the layers immediately above and below, and can talk to them regardless of departmental boundaries.

The Japanese compare strong vertical and horizontal linkages to a well-constructed piece of fabric. For a good weave, both the vertical and horizontal threads must be strong (see Figure 4–3). Although the Japanese also have line organizations, cross-functional activities are strengthened through the use of QFD.

Before QFD can begin, the customer must be identified. In most instances, more than one customer exists—for instance, the end user, the company the product is being produced for, and the assembly operator who will be putting the product together. In almost all cases, there will be both external and internal customers. Both types need to be taken into account; but should a conflict arise, the internal customers should almost always take a backseat to the external customers, thus ensuring that end users get what they want.

QFD is accomplished through a series of charts and matrices that may seem very complex at first glance. When broken into its individual elements, QFD isn't difficult to understand. The process simply emphasizes what needs to be done and how to go about doing it.

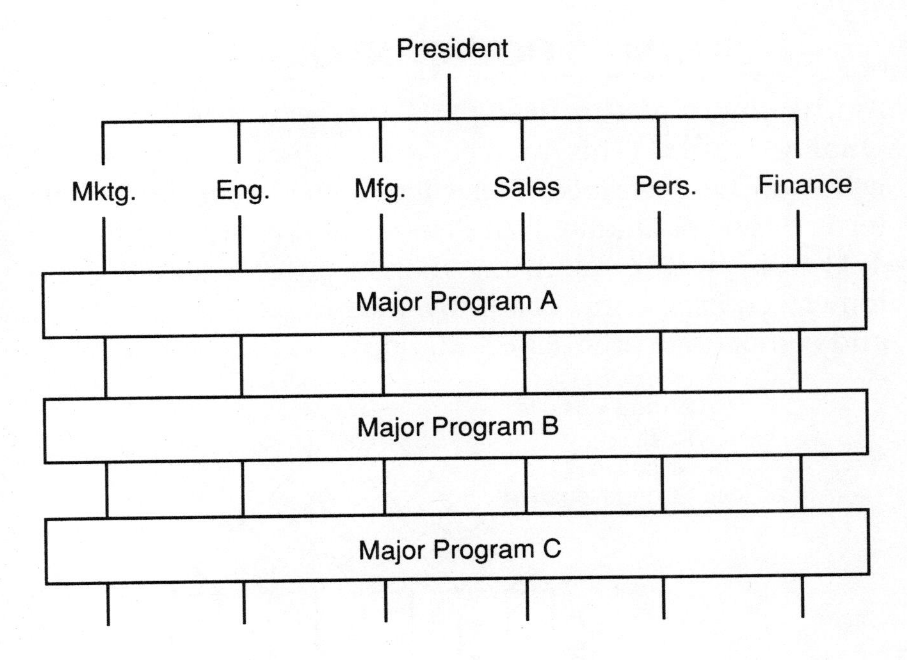

FIGURE 4–3
QFD helps strengthen both vertical line organizations and horizontal program linkages, which enhances the product development process.

For practical purposes, QFD can be thought of as a four-part process: Phases one and two address product planning and product design, and phases three and four address process planning and shop floor activities. In actuality, however, many companies adapt the four-part model to their particular needs. For example, ITT Mechanical Systems and Components successfully used a three-part QFD procedure to shorten the time needed to implement the methodology. Companies in the service industry, on the other hand, typically skip phase two, whereas companies involved in continuous batch-processing endeavors might combine phases two and three.

BUILDING A HOUSE OF QUALITY

At the heart of the first QFD phase is the House of Quality matrix. (This matrix's correlation matrix, which will soon be described, looks like a tiled roof, hence the term House of Quality.) The House of Quality (see Figure 4–4) is a product-planning matrix used to depict customer requirements, company measures, target values, and competitive product evaluations.

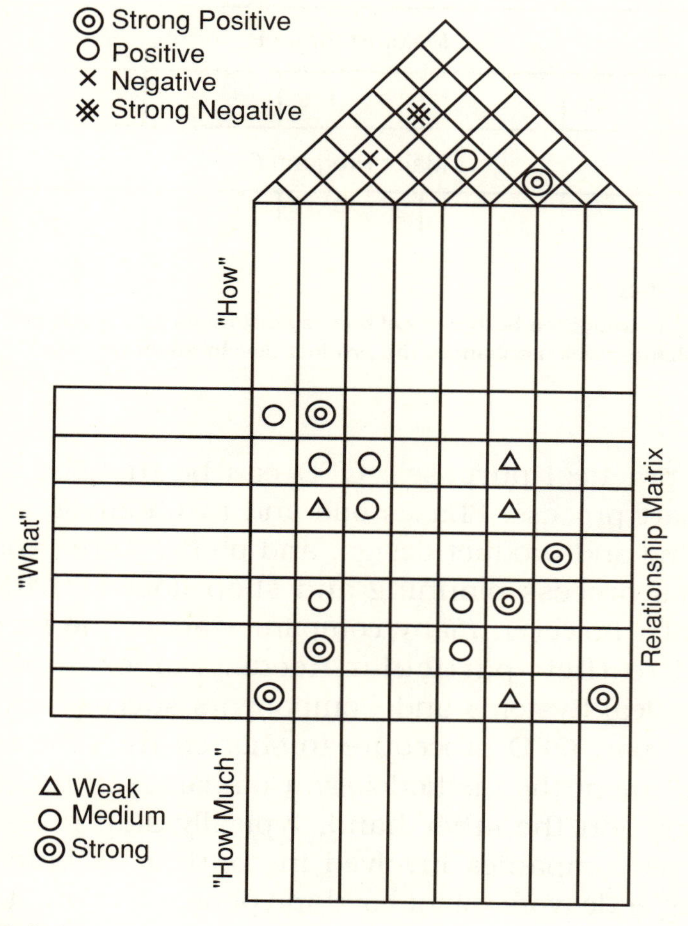

FIGURE 4–4

The House of Quality product planning matrix charts customer requirements, company measures, target values, and competitive product evaluations.

The following summary of the House of Quality components will help clarify the content and function of each element. Touring the rooms of the House of Quality will provide better understanding the house itself. But first, examine a recurring QFD theme: from *what* to *how to how much* (see Figure 4–5).

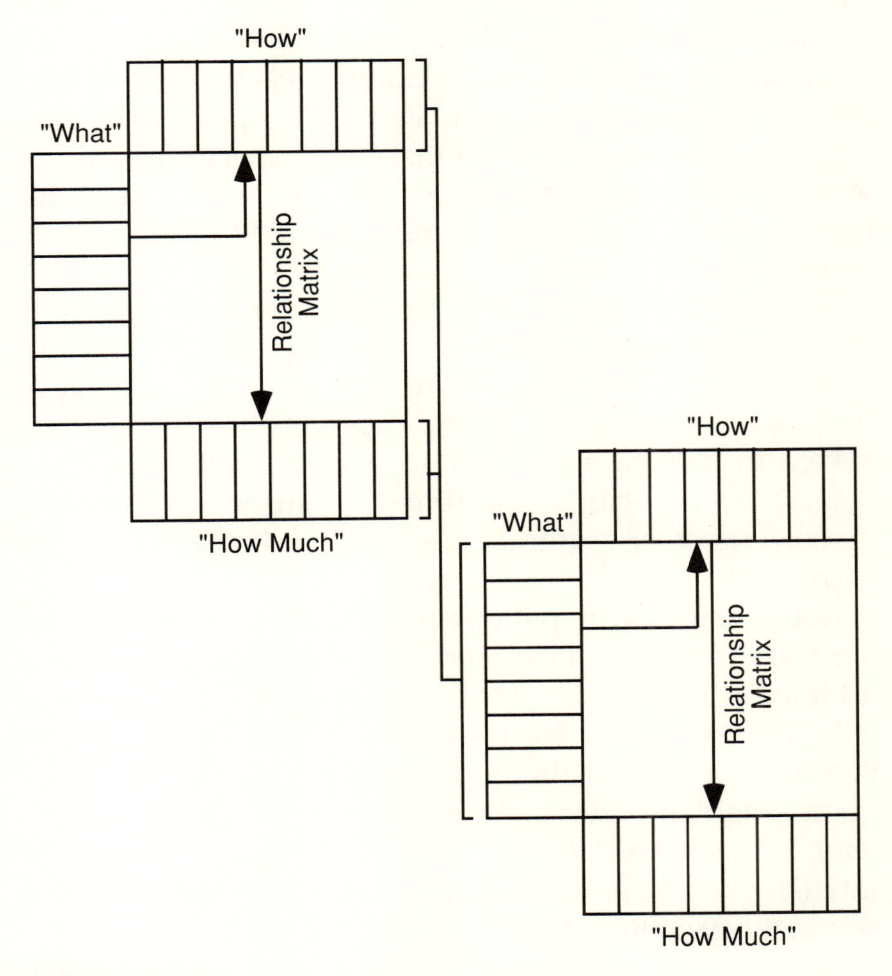

FIGURE 4–5

A *what* to *how* to *how much* theme is common to most QFD matrices and charts. This theme is based on an input-output strategy: *What* items are broken into *how* items, and then the *how* items evolve into new *what* items.

This theme is loosely based on an input-output strategy. QFD begins with a list of loosely stated objectives—the *whats* that should be accomplished. These *what* items are the basic customer requirements. They will probably be vague and require further detailed definition. One such *what* item might be an excellent cup of coffee (see Figure 4–6). Every coffee-drinker wants this, but providing it requires further definition.

To provide the required definition, each *what* item is broken into one or more *how* items. This process is similar to the process of refining marketing specifications into system-level engineering specifications. Customer requirements are actually being translated into company measures.

The excellent cup of coffee requirement, for example, would be translated into "hot," "eye-opener," "rich flavor," "good aroma," "low price," "generous amount," and "stays hot" (see Figure 4–7). If the cup of coffee were being served in a restaurant, "service with a smile" and "free refills" might also be customer requirements or *how* items. Customer requirements and conditions of use are correlated.

How items also usually require further definition, and are thus treated as new *what* items that are broken into additional *how* items. This is similar to the process of translating system-level specifications into parts-level specifications. With the cup of coffee example, new *how* items might be "serving temperature," "amount of caffeine," "flavor component," "aroma component," "aroma intensity," "sale price," "volume," and "temperature after serving" (see Figure 4–8).

This refinement process is continued until every item on the list is actionable. Such detail is necessary because

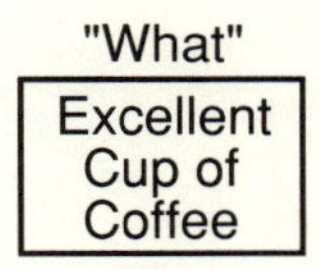

FIGURE 4–6

***What* items are the basic customer requirements—in this example, an excellent cup of coffee.**

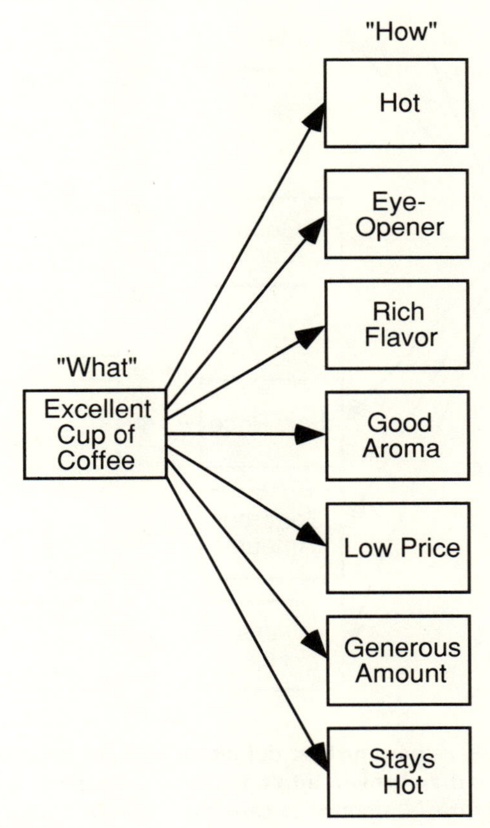

FIGURE 4–7

***What* items usually require further definition and are broken into one or more *how* items—in this example, "hot," "eye-opener," "rich flavor," "good aroma," "low price," "generous amount," and "stays hot."**

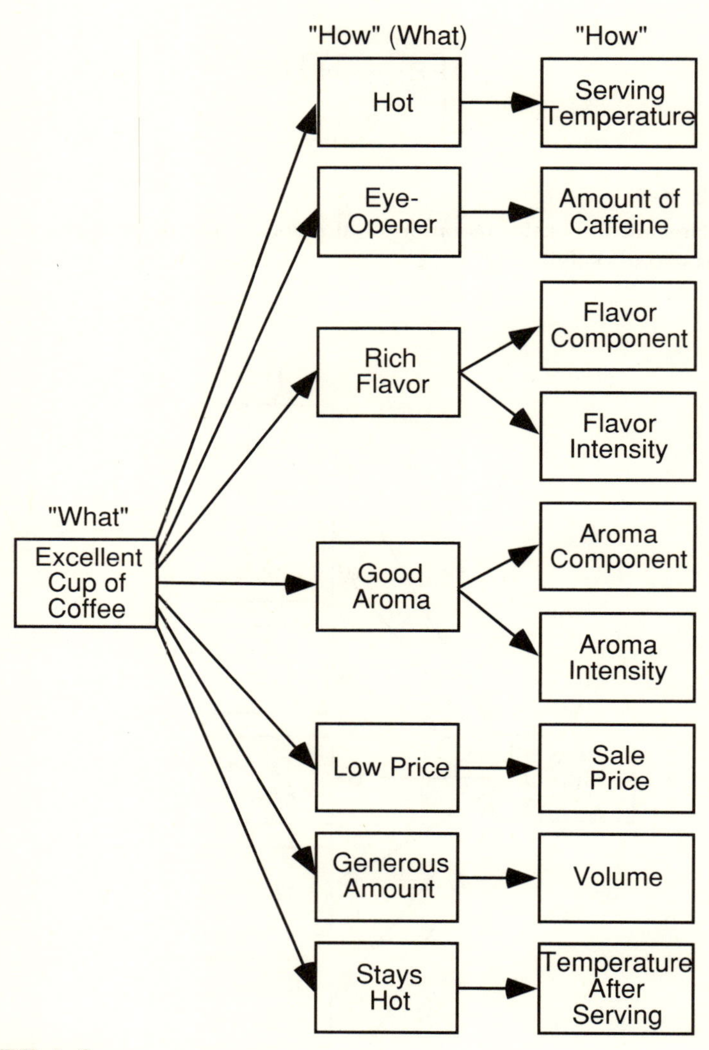

FIGURE 4–8

***How* items also usually require further definition and are treated as new *what* items. These are then broken into additional *how* items—in this example, "serving temperature," "amount of caffeine," "flavor component," "flavor intensity," "aroma component," "aroma intensity," "sale price," "volume," and "temperature after serving."**

there is no way of successfully performing a requirement that no one knows how to accomplish!

Unfortunately, however, this process is complicated by the fact that some of the *how* items affect more than one *what* item, and can even affect one another. Only about half of all product improvement efforts are effective; the remaining 50 percent either fail to provide the desired improvement or introduce some unexpected problem. This occurs with even the best engineers, because such complex relationships cannot be fully comprehended.

Attempting to clearly trace the relationships of *what* and *how* items becomes quite confusing at this point (see Figure 4-9). QFD provides a way of untangling this complex web of relationships via a matrix bordered by *how* and *what* items that defines the relationships.

The customer-defined *what* items are listed to the left of the relationship matrix on a vertical axis. The *how* items (company measures) are listed on a horizontal axis above the relationship matrix (see Figure 4–10).

Next, relationships of the *what* and *how* items are symbolically represented: Unique symbols are used to depict weak, medium, and strong relationships between customer requirements and company measures. Commonly used symbols are a triangle for weak relationships, a circle for medium relationships, and a double circle for strong relationships.

For example, customer requirements (*what* items) identified for an excellent cup of coffee were hot, eye-opener, rich flavor, good aroma, low price, generous amount, and stays hot. Corresponding company measures (*how* items) were serving temperature, amount of

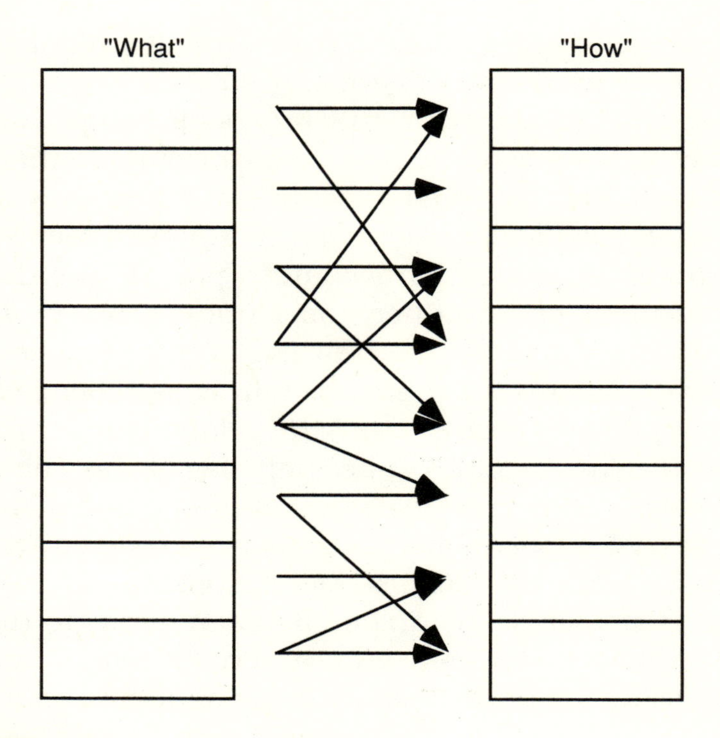

FIGURE 4–9

Clearly tracing the relationship of *what* and *how* items can be confusing—some *how* items affect more than one *what* item and can even affect one another. The House of Quality matrix simplifies this process.

caffeine, flavor component, flavor intensity, aroma component, aroma intensity, sale price, volume, and temperature after serving. The strong relationships are represented by double circles. These include the following: hot and serving temperature; rich flavor and flavor component; good aroma and aroma component; low price and sale price; generous amount and volume; and stays hot and temperature after serving. Medium relationships are depicted by circles. These are: hot and temperature after serving; eye-opener and serving temperature; eye-opener and amount of caffeine; rich flavor and flavor intensity; good aroma and

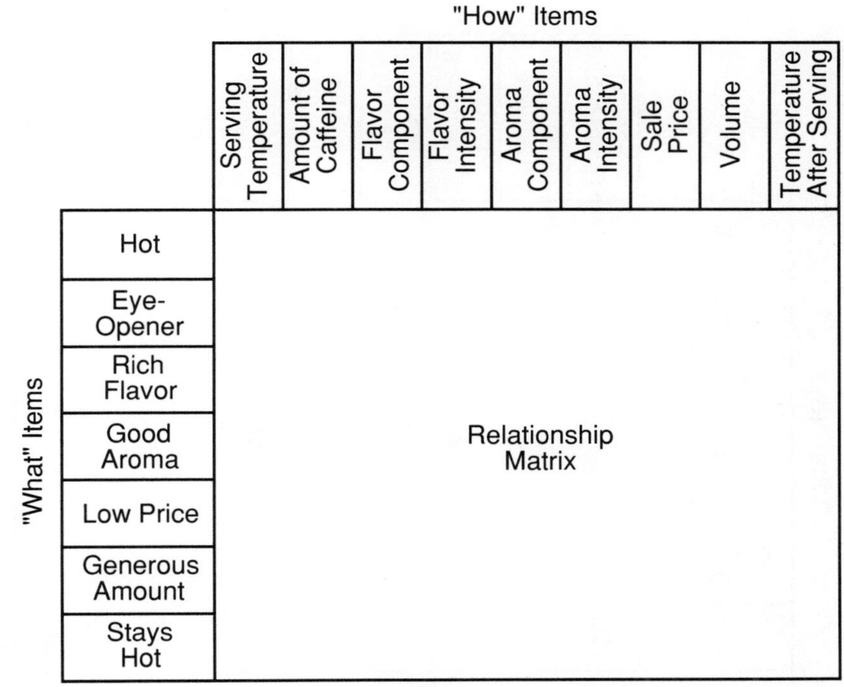

"How" Items

"What" Items	Serving Temperature	Amount of Caffeine	Flavor Component	Flavor Intensity	Aroma Component	Aroma Intensity	Sale Price	Volume	Temperature After Serving
Hot									
Eye-Opener									
Rich Flavor									
Good Aroma					Relationship Matrix				
Low Price									
Generous Amount									
Stays Hot									

FIGURE 4–10
The *what* items are listed to the left of the House of Quality's relationship matrix on a vertical axis. The *how* items are listed on a horizontal axis above the relationship matrix.

aroma intensity; low price and volume; generous amount and sale price; and stays hot and serving temperature. Weak relationships are shown as triangles. These are: rich flavor and serving temperature and rich flavor and amount of caffeine. See Figure 4–11.

If no relationship exists, the matrix space is left blank. Blank rows or blank columns indicate places where the translation of *what* items into *how* items is inadequate, providing an opportunity for valuable cross-checking. The QFD process provides numerous opportunities for cross-checking, which is one of its greatest strengths.

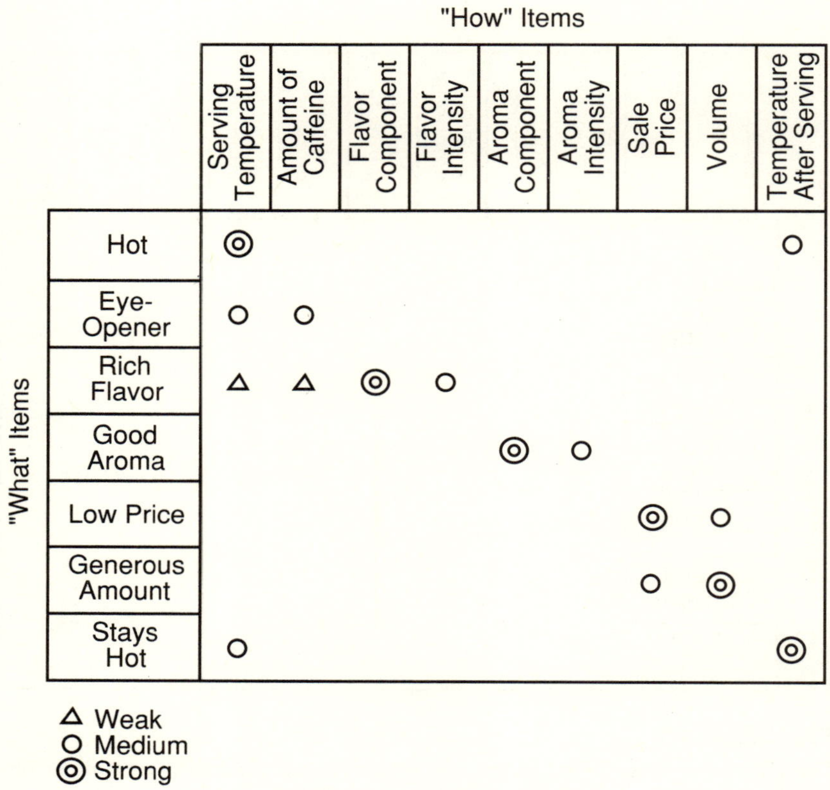

"How" Items

"What" Items	Serving Temperature	Amount of Caffeine	Flavor Component	Flavor Intensity	Aroma Component	Aroma Intensity	Sale Price	Volume	Temperature After Serving
Hot	◎								○
Eye-Opener	○	○							
Rich Flavor	△	△	◎	○					
Good Aroma					◎	○			
Low Price							◎	○	
Generous Amount							○	◎	
Stays Hot	○								◎

△ Weak
○ Medium
◎ Strong

FIGURE 4–11
Unique symbols are used to depict the relationships between customer requirements (*what* items) and company measures (*how* items). These particular symbols define weak, medium, and strong relationships.

The ability of QFD to evolve plans into actions while repeatedly cross-checking the thinking process also makes it useful for any nontrivial planning function.

Running parallel with the *how* axis on the bottom edge of the relationship matrix is a third element, the *how much* axis. *How much* items are measurements for the *how* items. *How much* items are kept separate from the *how* items because when the *hows* are determined, the values of the *how much* items usually aren't known. These values will be determined through analysis.

How much items provide both an objective means of ensuring that requirements have been met and targets for further detailed development. Thus, *how much* items provide specific objectives that guide the subsequent design and afford a means of objectively assessing progress. Whenever possible, *how much* items should be measurable. Measurable entities tend to be actionable. They provide more opportunity for analysis and optimization than nonmeasurable entities. Without measurements to work toward, the goal is often merely to do better. On the other hand, what gets measured gets improved. If most of the *how much* items aren't measurable, *how* item definitions probably aren't detailed enough.

How much items for the cup of coffee example include 120–140 degrees Fahrenheit, —— ppm, less than $.25, greater than 12 fluid ounces, and 110–125 degrees Fahrenheit (see Figure 4–12). *How much* items related to the flavor and aroma customer requirements are determined by a panel of judges.

The *what/how/how much* process forms the basis for almost all QFD charts. It's the key that unlocks the House of Quality. Enhancing this process are the correlation matrix, competitive assessment graphs, and rating and weighing.

Correlation Matrix

The triangular correlation matrix, the roof of the House of Quality and hence the source of its name, is located parallel to and above the *how* axis (see Figure 4–13). This matrix describes the correlation between each *how* item

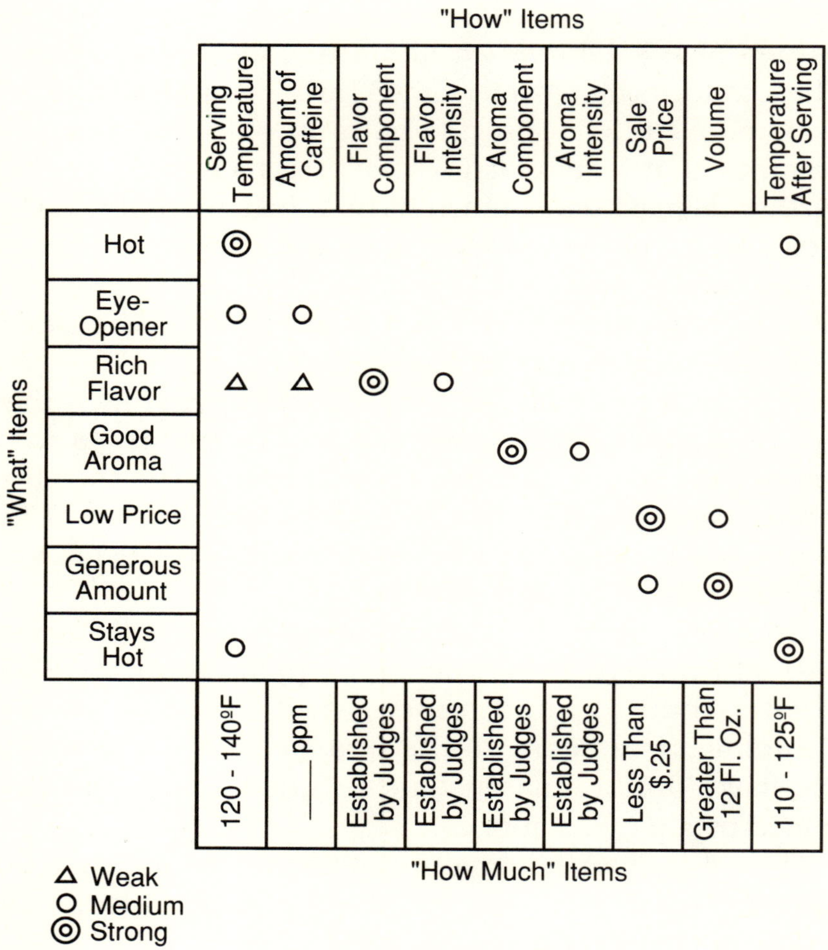

"How" Items

"What" Items	Serving Temperature	Amount of Caffeine	Flavor Component	Flavor Intensity	Aroma Component	Aroma Intensity	Sale Price	Volume	Temperature After Serving
Hot	◎								○
Eye-Opener	○	○							
Rich Flavor	△	△	◎	○					
Good Aroma					◎	○			
Low Price							◎	○	
Generous Amount							○	◎	
Stays Hot	○								◎
"How Much" Items	120 - 140ºF	____ ppm	Established by Judges	Established by Judges	Established by Judges	Established by Judges	Less Than $.25	Greater Than 12 Fl. Oz.	110 - 125ºF

△ Weak
○ Medium
◎ Strong

FIGURE 4–12

***How much* items (for example, 120–140° F, less than $.25, greater than 12 fl. oz., and 110–125° F), measurements for the *how* items, are listed at the bottom edge of the House of Quality's relationship matrix on a horizontal axis.**

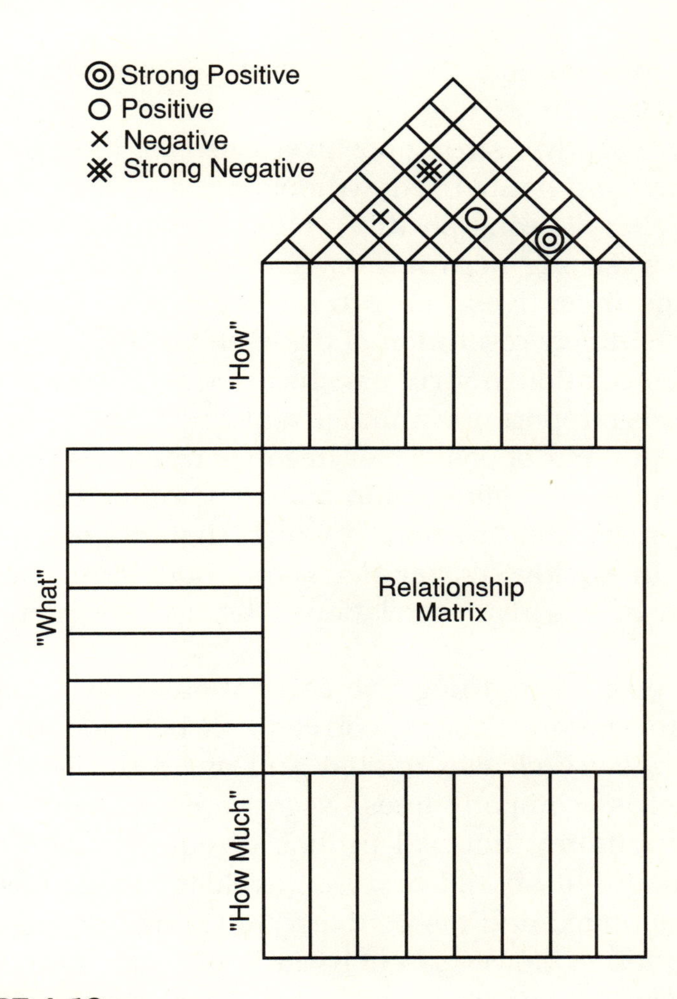

FIGURE 4–13

The correlation matrix, which describes the correlation between each *how* item via unique symbols that represent positive or negative ratings and the strength of each relationship, is located parallel to and above the *how* axis.

via unique symbols that represent positive or negative ratings and the strength of each relationship (that is, positive, negative, strong positive, or strong negative correlation). Commonly used symbols are a circle (positive), double circle (strong positive), cross (negative), and double cross (strong negative). By charting conflicting relationships (negatives and strong negatives), the matrix facilitates timely resolution of trade-off issues.

The correlation matrix can be used to identify which *how* items support one another and which are in conflict. The assignment of positive or negative ratings is based on each *how* item's influence in achieving other *how* items, regardless of the direction in which the *how much* value moves. In positive correlations, one *how* item supports another. In negative correlations, the two *how* items are in conflict.

Both positive and negative correlations provide important information. Positive correlations help identify *how* items that are closely related and avoid duplication of effort across company lines. Negative correlations represent conditions that will probably require trade-offs—conditions that should never be avoided. Trade-offs that aren't identified and resolved lead to unfulfilled customer requirements; however, such trade-offs can usually be easily resolved by adjusting the *how much* values.

Competitive Assessment

Two competitive assessment graphs (see Figure 4–14) provide an item-by-item comparison between a company's product and similar competitive products. The first of these graphs, listed on a vertical axis to the right of the relationship matrix, corresponds to the *what* items, and the second graph, listed on a horizontal axis to the right of the relationship matrix, corresponds to the *how* items.

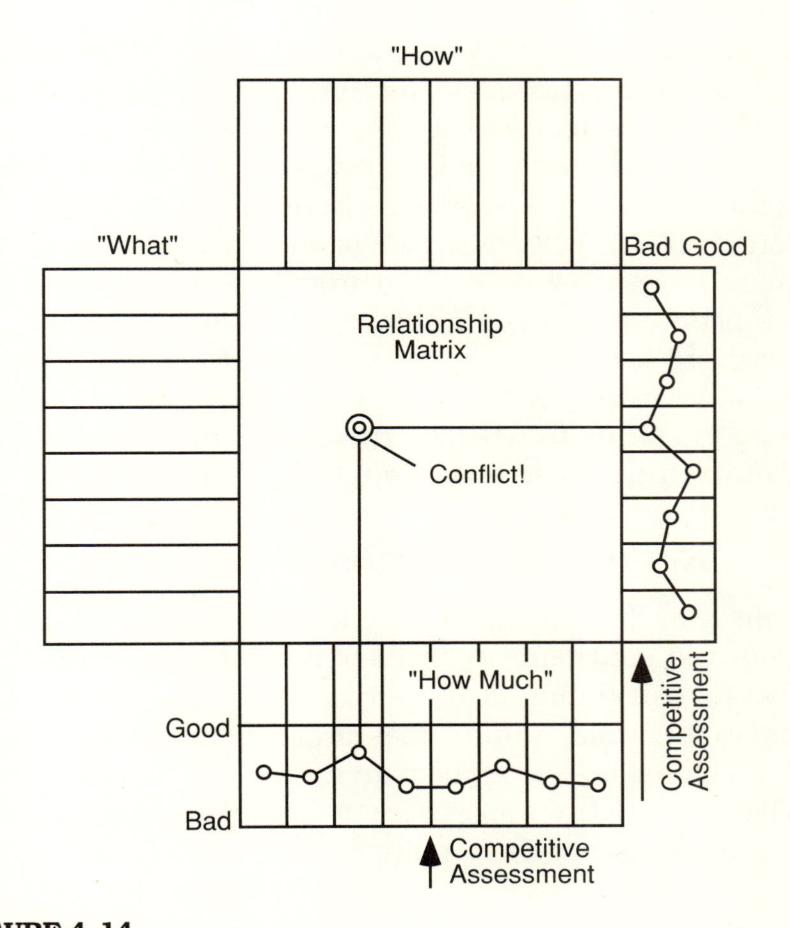

FIGURE 4–14

Competitive assessment graphs, which provide an item-by-item comparison between a company's product and similar competitive products, are listed on a vertical axis to the right of the House of Quality's relationship matrix and on a horizontal axis below the relationship matrix.

Competitive assessment of the *what* items is also called customer competitive assessment and should use customer-oriented information. Technical competitive assessment, which should use engineering-generated information, is another name for competitive assessment of the *how* items.

The competitive assessment graphs can be used to establish competitive *how much* values and help position a product in the marketplace. They are also extremely beneficial when used to detect gaps or errors in engineering judgment—including instances where in-house evaluations don't coincide with the voice of the customer. If the *how* items have properly evolved from the *what* items, their competitive assessments should be similar. Strongly related *what* and *how* items should also exhibit a similar competitive assessment relationship. Thus, these graphs provide a cross-check as to whether the *hows* are the right *hows*, allowing the team to move forth with greater confidence.

Rating and Weighting

Charts of graphs that numerically rate and weight the *what* and *how* items in terms of the desired end result are also useful. Rating of the *what* items is performed on a one-to-five scale. A numerical rating of one to five is placed in a column to the immediate right of each *what* item to reflect the relative importance of this item to the customer. These ratings are then multiplied by the weights assigned to each matrix symbol (that is, weak, medium, and strong).

The standard 9-3-1 weighting system is often used, although alternative systems can be applied to the same effect, placing stronger emphasis on the most important items. The results of the rating and weighting exercises are then recorded on a horizontal axis below the *how much* items (see Figure 4–15). This results in the identification of critical product requirements, which translate to critical customer requirements, and aids in the trade-off decision-making process.

This example of a House of Quality is just that—an example. Houses of Quality can be built in many shapes and forms to meet almost any need. There is even a

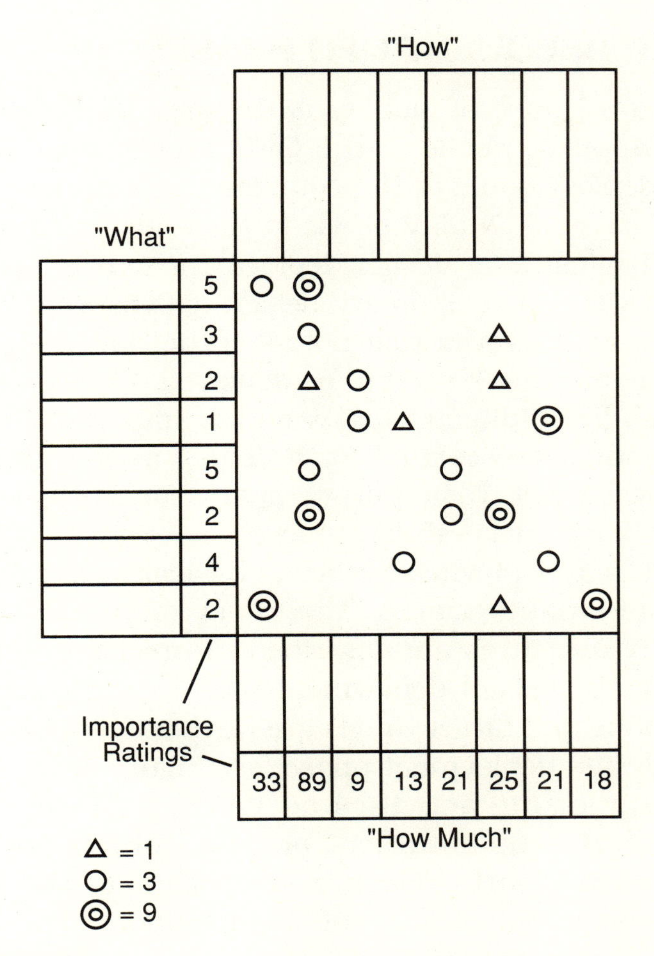

FIGURE 4–15

Charts or graphs that numerically rate and weight the *what* and *how* items are located to the right of the vertical *what* axis and below the horizontal *how much* axis.

Condo of Quality discussed in chapter 5. The important thing is that Houses of Quality be tailored to the individual application. Some of the additional elements that can be included are key selling points, the level of technical difficulty, technical standards, and quality standards.

SUBSEQUENT QFD PHASES

Building a House of Quality is the first, and most commonly applied, phase of the QFD procedure. The next phase deploys some of the company measures identified in the House of Quality phase to the subsystem or parts level. The resulting design deployment matrix serves as the basis for all preliminary design activities. It's important to note, however, that not all of the House of Quality company measures need to be deployed. Rather, only the high-risk (new, difficult, or extremely important) company measures are carried forth. This ensures that time and effort aren't wasted on company measures that are already being successfully achieved.

The design deployment matrix physically resembles the House of Quality matrix. Customer requirements and company measures are described in precise engineering terms in the design deployment matrix, and competitive evaluations and target values are further developed.

The design deployment phase (see Figure 4–16) utilizes such supporting activities as VAVE, FTA and RFTA, FMEA, design optimization, process optimization, cost analysis, and parts selection for reliability assurance. This phase culminates in the identification of part characteristics that are critical to the execution of the company measures. The critical part characteristics are highlighted in a design deployment chart that, in turn, helps identify manufacturing operations.

The process planning phase (see Figure 4–17) represents the transition from design to manufacturing operations. A process planning chart is prepared for each critical part characteristic at this phase.

The following information is included in each process planning chart: a list of the required processes, a matrix

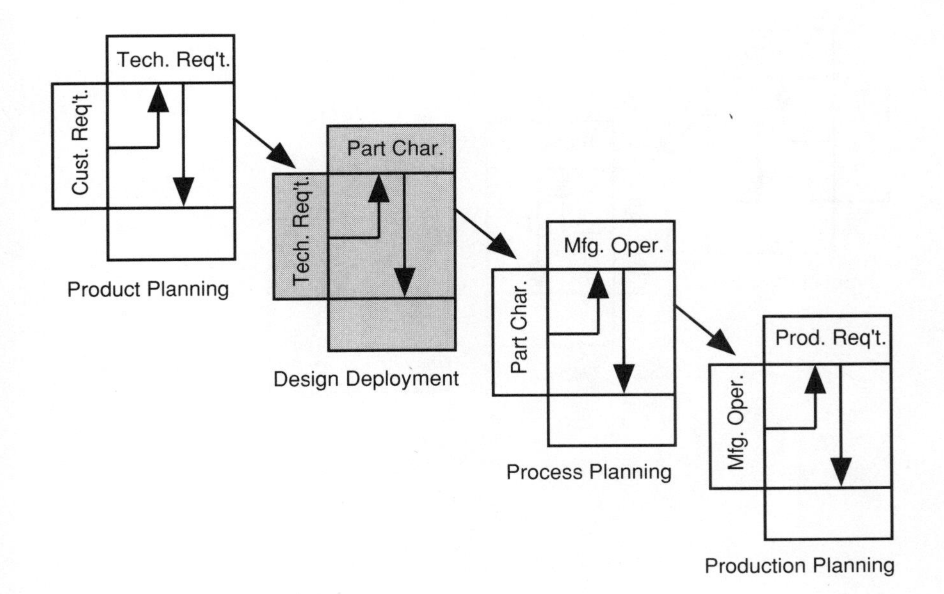

FIGURE 4–16

The design deployment QFD phase deploys the company measures from the House of Quality phase to the subsystem or parts level.

graphing the relationship between each process and each critical part characteristic, and lists of process control parameters. This information is used to produce process control charts for each part. During this phase, a process FMEA is conducted, and the information from the previous charts is verified and reviewed.

The production planning phase (see Figure 4–18) transfers the information generated in the subsequent phases of QFD to the factory floor via a series of tables and charts. This phase deploys information relevant to a number of functions. Like the other QFD phases, it can be adapted to meet a wide variety of requirements. Obviously, QFD thrives on flexibility.

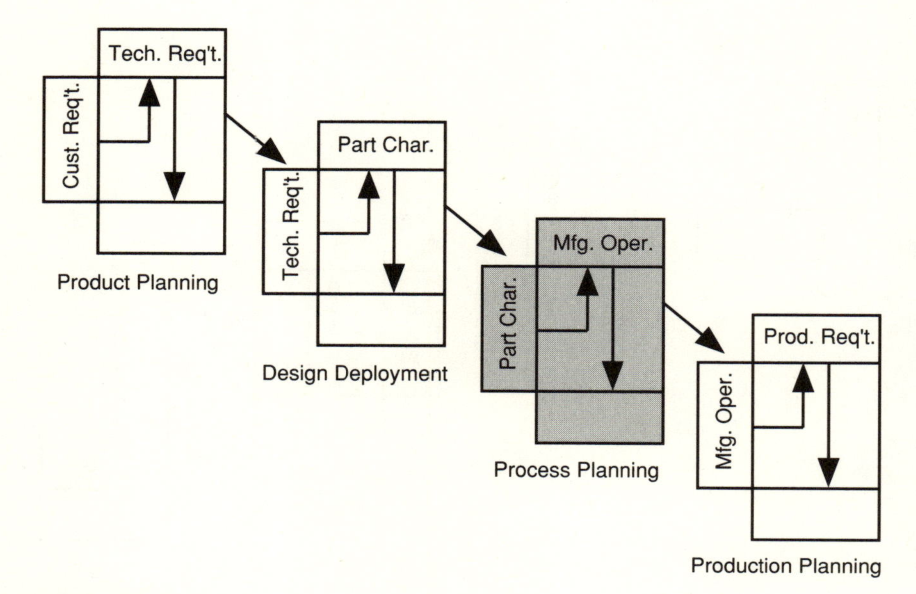

FIGURE 4–17
The process planning QFD phase represents the transition from design to process planning.

PUTTING IT ALL TOGETHER

As noted earlier in this chapter, the QFD process begins with customer requirements and then translates them into technical requirements, part characteristics, manufacturing operations, and production requirements. The opportunity to reduce variability extends throughout the process. Variability is not only the enemy of quality, it is the enemy of profitability and productivity as well. Without variability, products, processes, and systems behave in an extremely consistent manner. Due to variability, however, they don't always behave the way they're supposed to; hence, the troublesome trio of scrap, rework, and customer problems.

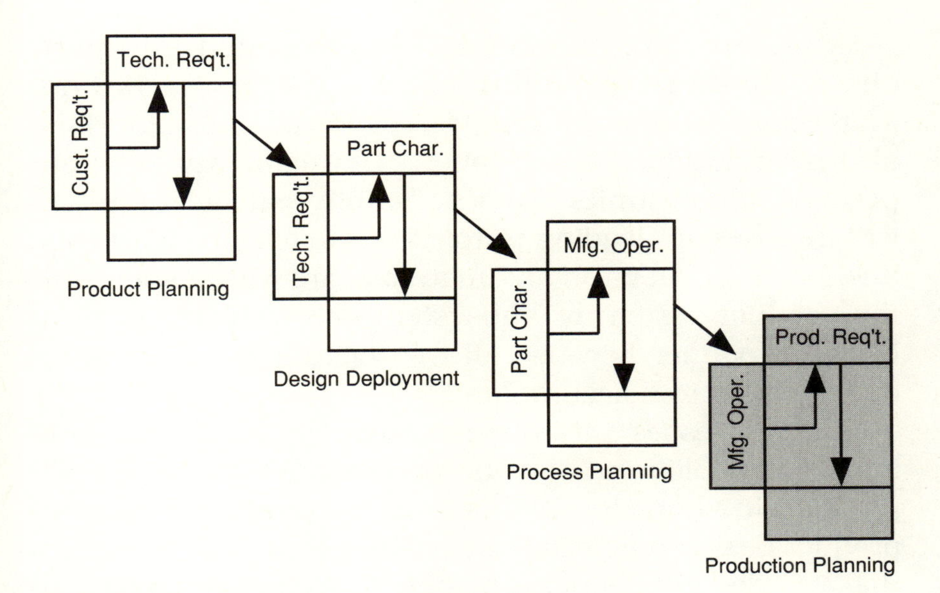

FIGURE 4–18
The production planning QFD phase transfers the information generated in the subsequent QFD stages to the factory floor.

Variability is most often seen in manufacturing, the focus of most variability reduction efforts. In actuality, many causes of variability go far beyond manufacturing. The QFD process provides an opportunity to address variability throughout the entire product development process, and if done properly, manufacturing will not have to contend with much variability.

As customer requirements are translated into technical requirements, the variability between what the customer wants and product specifications can be substantially reduced. When this isn't the case, the resulting product may be the world's best Edsel—clearly meeting specifications but failing to satisfy the customer.

As technical requirements are translated into part characteristics, variability contributed by parts and materials can also be minimized. Parts and materials also have assigned specifications that can vary between predetermined limits. This variability can accumulate within a design, leading to tolerance stack up. With this phenomenon, all parts and materials are within spec, but the product system or subsystem doesn't perform in a likewise manner because all the tolerances stack up in one direction and accumulate.

QFD addresses this phenomenon by minimizing the impact variability has on the performance of the overall product. In short, the goal is a robust product design—a product design reasonably insensitive to the many things that will degrade its performance, including variability in parts and materials. (For more on achieving robust design at low cost, see the appendix on Taguchi Methods.)

As part characteristics are translated into manufacturing operations, the influence of variability from the manufacturing process, equipment, or environment is minimized. Both the process and the environment will contribute some variability, the reduction of which is very costly. For example, a petroleum refinery is subject to variations in ambient temperature, a condition that could be changed by enclosing the entire refinery in a climate-controlled building. Unfortunately, the cost to do so would be prohibitive. A better option would be to create a robust process design—a process design reasonably insensitive to the many factors that tend to deteriorate its performance, such as the environment and variability in machinery and operating techniques.

Finally, as manufacturing operations are translated into production requirements, the variability that occurs due to people needs to be minimized. This is achieved by using the most effective operating practices on the shop floor.

In addition, QFD is often used in conjunction with *simultaneous* or *concurrent engineering.* This is a systematic approach to product design development that considers all elements of the product life cycle, from conception through disposal, and thus simultaneously defines the product, its manufacturing process, and any additional processes, such as maintenance.

As noted by Stephen ter Haar and others (1993),

> QFD contributes to all three key elements of Concurrent Engineering. . . . It is integrated within a multifunctional team approach, it strongly supports many decision-making processes throughout the whole development project, and it is integrative in the sense of connecting all these decision-making processes.

MIT's Don Clausing agrees that incorporating QFD and concurrent engineering strategies makes a lot of sense. Clausing has coined the term *enhanced QFD* to describe the integration of QFD and Pugh Concept Selection, which uses a comparison-evaluation matrix to compare available concepts against established criteria for evaluation drawn from the House of Quality. According to Clausing, both processes are highly visual and connective. Thus, they work extremely well together.

And ter Haar, who studied with Clausing at MIT, has combined QFD and the Design Structure Matrix (DSM), another technique that expresses design procedure information in a compact, visual form. According to ter Haar (1993),

> QFD is considered an essential part of Concurrent Engineering, and it is in widespread use in various industries. Unlike the Design Structure Matrix, the initial planning tool of QFD, the House of Quality, is fed by the voice of the customer. Thus, the following planning processes within QFD are inherently responsive to the needs of the customers as well. . . . Due to the

attributes of QFD one might expect that the information generated during the QFD procedure can be used to update the DSM, which is based on comparable previous design processes.

QFD has also been applied in conjunction with a variety of other processes and methodologies. As noted in chapter 3, the QFD methodology can help an organization determine the most effective application for any number of engineering and analytical tools, including the cause-and-effect diagram, competitive benchmarking, FMEA, FTA, Pareto Analysis, Pugh Concept Selection, and Taguchi Methods (see Figure 4–19). QFD is more than a quality tool. It is a planning methodology that brings out the best in a palette of potentially useful quality tools and techniques by clearly painting when and where to use them.

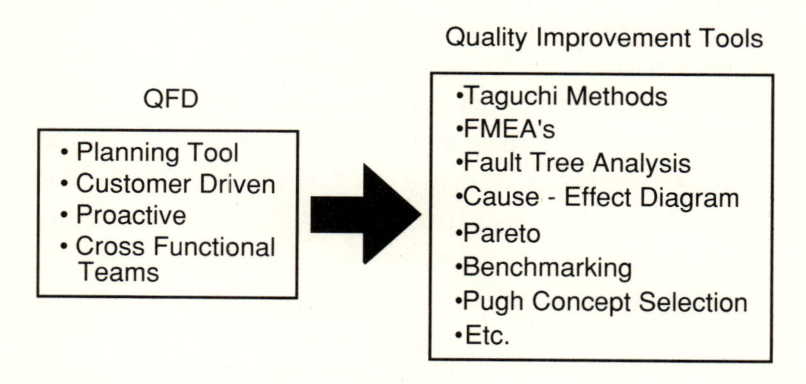

FIGURE 4–19
QFD brings out the best in a wide variety of quality tools and techniques by identifying when and where to use them.

QFD at P&G: "Seek First to Understand"

Earl C. Conway is the retired corporate director of quality worldwide for Procter & Gamble and adjunct professor of industrial engineering at the University of Cincinnati. He quoted Stephen Covey, author of *The Seven Basic Habits of Highly Effective People*, when asked about QFD: "Seek first to understand before you ask to be understood."

An acclaimed lecturer and consultant who frequently worked with the late quality improvement pioneer Dr. W. Edwards Deming, Conway (1993) warns of the dangers involved when QFD is taught by "consultants and training organizations that have a fundamental flaw in their understanding—often leading to what Dr. Deming called 'hacks teaching hacks.'"

Conway introduced QFD to consumer goods giant P&G in 1986, strategically arranging an executive briefing for the top 50 officers of the company. A year later, QFD was also presented to the head of P&G's international operations, resulting in the introduction of the methodology to P&G's general managers in Europe, Asia, and Latin America.

"These were key moves," Conway explains. "QFD first must be understood and supported in principle by the top layers of a company. In retrospect, it was a casebook study of how to do it right. If we hadn't first involved our top management, we wouldn't have been nearly as successful."

Conway acknowledges that introducing QFD to top management is not easy, with executive briefings taking a good half day away from executives' busy schedules. But in the case of QFD, the time spent quickly becomes time saved.

According to Conway, it is essential that top management, not middle operations management, first be exposed to QFD. He notes the potential misunderstanding involved when "senior managers observe their people spending so much time on those seemingly complex charts that they [senior management] know nothing about."

Conway describes QFD as "one of the most important product design planning tools in use at P&G," a methodology that has been successfully applied to both new product design and the improvement of existing products. For example, Pampers and Luvs disposal diapers' recent design improvements all evolved from early QFD studies, as did the Crest Complete toothbrush, a new product for P&G. Although the company had manufactured toothpaste brands (Crest and Gleem) for years, it saw no opportunity to enter the toothbrush market with a product design advantage until applying QFD.

In addition, the methodology is also frequently applied to system or process design by the worldwide market leader; for example, in the design of product distribution systems and even in the creation of advertising copy.

Noting that P&G has trained "hundreds of people in QFD" and "has had as many as 100 QFD teams working at the same time in the Cincinnati area alone," Conway concedes that P&G ran into the occasional obstacle when first implementing QFD. P&G brought in experts from the American Supplier Institute and Akasha Fukuhara of the Central Japan Quality Control Association until the P&G teams became clearly self-sufficient.

Conway advises that potential users also can be helped to understand what QFD *is* by first understanding what it is *not.* "QFD is not a replacement for fundamental and ongoing customer research," he says. "On the other hand, QFD can help alert you when you have the wrong data, not enough data, or too much data. It can also help you realize when your project no longer has value."

Although Conway notes that QFD should play a major role in improving the competitiveness of U.S. industry, he warns that the methodology should only be applied to complex design projects where there is a major need—make that a major opportunity—for improvement.

Customer-Driven Design

The result of the first QFD case study performed at the Kelsey–Hayes Company in Romulus, Michigan, was a customer-driven design for a new electromechanical product, a coolant-level sensor. The study was performed in conjunction with the American Supplier Institute and Ford Motor Company, for which the sensor was being produced.

The QFD study was composed of three parts: (1) assembly and component features, (2) material and design, and (3) manufacturing processes. The market quality (customer) requirements were broken into functional and performance requirements. Functional requirements were then broken into customer and operation categories, and performance requirements were broken into serviceability and durability.

After completion of an initial House of Quality for the coolant-level sensor, the housing was identified as the most critical product (design) requirement. A House of Quality was then completed for the housing itself. Key customer functional and performance requirements included easy-to-add coolant, easy-to-identify unit, and provide cap removal instructions.

As a result of the QFD study, a coolant-level sensor was designed with product features that would best meet the stated functional and performance requirements. Included in the final production design, for example, were slotted holes in the tube to allow coolant to freely flow to the coolant reservoir. Also, the words *Radiator Coolant Only* in raised yellow lettering and a single tab labeled *Lift* were added to the housing cap.

The QFD study was beneficial in more ways than one. "As an engineer on that first project, I'm able to say that doing a QFD study made me aware of the product in terms of how it would fulfill customer requirements," explained Peter J. Soltis, then senior technical specialist of product engineering at Kelsey–Hayes.

REFERENCES

Clausing, D.P. 1988. Personal interview. Interviewed by Nancy Ryan.

Conway, E.C. 1993. Telephone interview. Interviewed by Nancy Ryan.

Covey, S.R. 1989. *The Seven Basic Habits of Highly Effective People: Powerful Lessons in Personal Change.* New York: Simon & Schuster.

"How to Regain the Productive Edge." 1989. *Fortune,* 22 May, 92-4+. This is a review of the MIT study *Made in America: Regaining the Productive Edge.*

ter Haar, S., D.P. Clausing, and S.D. Eppinger. 1993. *Intregation of Quality Function Deployment and the Design Structure Matrix.* Cambridge, Mass.:Massachusetts Institute of Technology.

Chapter Five

Customizing Your House

DIFFERENT STYLES OF HOUSES

After the customer requirements have been translated into company jargon, that is, company measures or engineering terminology, the first House of Quality can be built. But what if there are too many customer requirements and company measures to count? What if there are scores of these requirements, say 100 or so?

This many requirements cannot be managed in a traditional House of Quality. Luckily, however, other housing options exist.

Condos of Quality—the subsystem-level of Quality Function Deployments that help keep the project efficient and focused—can be used with a large number of customer voices. Managing a large number of voices can seem nearly impossible with only one matrix. The time requirements for determining the relationships and obtaining the supportive data are often overwhelming in such situations, not to mention discouraging.

Condos of Quality are better housing options when the number of voices exceeds 25, although up to 50 voices can be manageable with the traditional House of Quality, depending on the team and the project.

A preplanning matrix is usually used initially to prioritize and identify which issues to attack. This often leads to subsystem QFD studies, that is, Condos of Quality. To put this another way, if the product is as simple as a lawn sprinkler, a detailed QFD study could be done on it. But if the product is more complex, it is better to use a subsystem approach and the preplanning matrix, with the following exceptions: (1) when the product is so simple that a preplanning matrix is not required and (2) when the priority decisions that the matrix would make have already been made by other means.

After the preplanning matrix has been carefully completed, work on the subsystem QFDs can begin (see Figure 5–1). The QFD process for the House of Quality described in chapter 4 applies to the Condo of Quality as

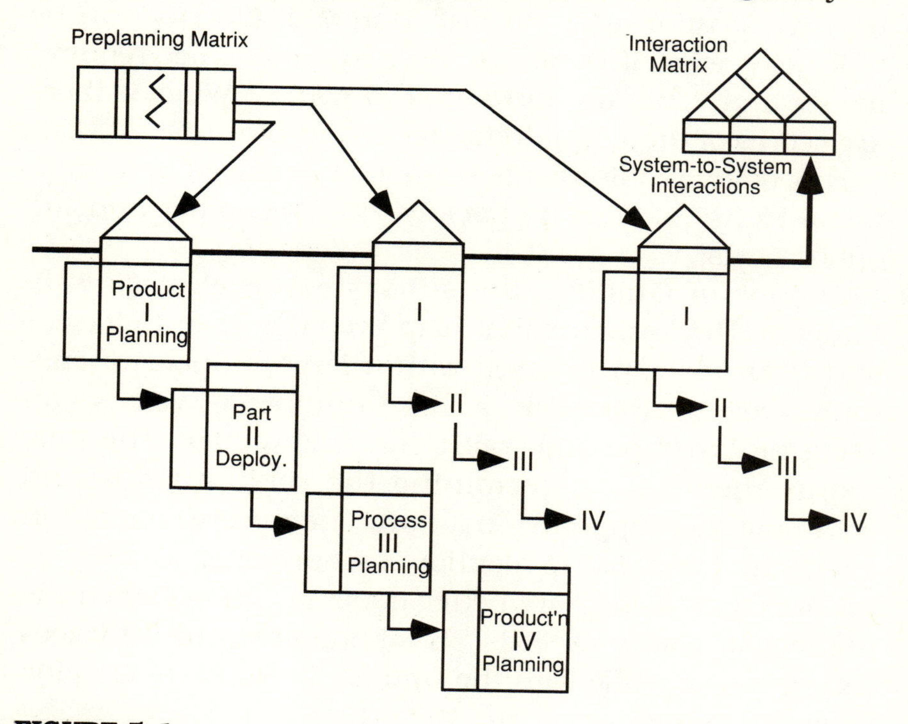

FIGURE 5–1

Complex products call for subsystem QFD studies, that is, Condos of Quality instead of Houses.

well, with one addition: An interaction matrix documents any system-to-system correlations (see Figure 5–2).

LIAISONS AND LINKAGES

QFD has perhaps the greatest potential for success when linked with a company's total business system; this includes marketing. Collaborating with marketing, however, can sometimes be problematic, because the marketing profession has an established tradition of how to do market research, and that tradition does not include QFD.

How does the market research that precedes the first stage of QFD differ from traditional marketing? To begin with, QFD is more of a qualitative approach, whereas traditional marketing is more quantitative. And although marketing people study the same problems, QFD has evolved from a completely different body of knowledge and may initially be viewed as a suspect activity. But despite some basic differences, both QFD and traditional

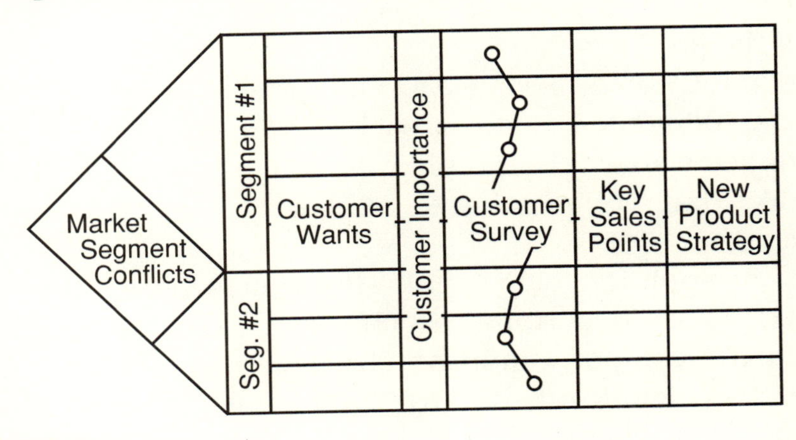

FIGURE 5–2
The interaction matrix documents system-to-system correlations that occur between subsystem QFD studies.

marketing research are working toward a common goal: a clear understanding of the voice of the customer. Thus, marketing people that see QFD in action typically buy into it. The proof is in the House of Quality.

Kevyn Irving (1993), a manager of technical services, contract development, used a skeleton in the closet approach to sell QFD to the marketing department and board of directors at Ethicon Endo–Surgery, a division of Johnson & Johnson based in Cincinnati. Due to the rapidly changing market demands in the booming field of endoscopic medical devices, time to market is a prime factor in Ethicon Endo–Surgery's business strategy.

In the traditional medical device field, a company can rapidly design a new product, move it into the marketplace, and drive the costs out while absorbing the losses. In the endoscopic medical device market, however, the product might very well be obsolete before the costs are driven out. Thus, new products are typically rushed to market in response to marketing department research but at significant expense. This expense can be devastating if the products fail.

As Irving noted, "People want to see results right away, versus taking the time to sit back and really see what the customer wants. The problem is, when you rush right out like that, you're going to get it wrong far more than get it right.

"In my opinion, QFD really flourishes here," Irving added before describing a QFD trial that could have saved the company months of development, retooling time, and significant funding. Interestingly enough, the trial QFD study identified 32 product features that were not in the revised product description, the majority of which are now considered important customer requirements. These include interchangeable shafts with attached end effectors and two different handle designs.

"This study added more credibility to our use of QFD," Irving noted. "There's a QFD chart in my office right now that's on its way to marketing. QFD's really getting into the system here." In addition, Ethicon Endo–Surgery is moving toward organizational changes that will enhance the team concept, an important aspect of QFD.

A technique coined Matrix Data Analysis can be employed during the marketing research stage of QFD. This tool, which represents the correlations found on the matrix chart on an *x* and *y* axis, is used to geographically portray market segmentation and to identify appropriate markets for both the user's company and the competition.

Although QFD can also be used for strategic and business planning, Policy Management (Policy Deployment) is the better choice. Policy Management is the process for developing achievable business plans and then deploying them throughout the company. It closely resembles QFD. The *what/how/how much* process used during a QFD study is nearly identical to that used to chart the goals and actions in Policy Management. The process results in a series of cascading matrices that closely resembles the House of Quality. See *The Process-Driven Business: Managerial Perspectives on Policy Management* for more on this subject.

While applying QFD to strategic or business planning will result in the development of individual plans and actions to accomplish corporate, divisional, and departmental strategies, Policy Management keeps a few extra items in its toolbox. These include the Plan-Do-Check Act (PDCA) Cycle and catchball.

QFD and Policy Management are definitely complementary and should be used in conjunction with one another whenever and wherever possible. First-tier Japanese companies, for example, often use Policy Management to solve their business planning needs and then apply QFD to carry their products through the development process. Several American companies, including Florida Power & Light and The Budd Company of Troy, Michigan, have also used QFD in conjunction with Policy Management.

To quote Norman E. Morrell (1988), corporate manager of quality product reliability at Budd, "Just as QFD translates the wants of the customer into actions that create a product, Policy Management ensures that your planning strategies are customer sensitive."

THE NEW QFD

The Fifth Symposium on Quality Function Deployment featured 45 case studies related to QFD, a dozen of which were pertinent to service-related and soft issues. The case studies covered a wide range, from "Applying QFD to Health Care Services: A Case Study at the University of Michigan Medical Center" and "The Application of QFD to the Los Angeles River Rescue Task Force" to "QFD and Personality Type: The Key to Team Energy and Effectiveness" and "QFD in Academia: Addressing Customer Requirements in the Design of Engineering Curricula."

The abstract for "Applying QFD to Health Care Services: A Case Study at the University of Michigan Medical Center," by Dr. Deborah M. Ehrlich and Dennis J. Hertz (1993) reads as follows:

> The University of Michigan Medical Center (UMMC) piloted Quality Function Deployment (QFD) in a new unit that consolidated several separated diagnostic procedures into one unit. Based upon early TQM successes, the organization employed QFD to realign resources to meet the valid customer requirements of

the combined groups in order to stimulate service volume by better satisfying customer desires. This paper will discuss the UMMC QFD approach, articulate experiences learned, identify changes that have been implemented, quantify the financial benefits that have resulted from these changes, and offer ideas on how to best utilize QFD at a referral hospital.

As a result of the increased interest in QFD for use in such diverse settings as hospitals and academia, both ASI and GOAL/QPC are offering increased training and course work on QFD for the service and administrative sector. The Quality Service Model featured in ASI's *QFD for Service Implementation Manual* is based on three key tracks: (1) plan for success, (2) improve the process, and (3) hold the gains. This conceptual model incorporates such QFD-dictated tools as the Kano model, preplanning matrix, House of Quality, Pugh Concept Selection, process flowcharts, and control plans. The conceptual model consists of the following steps.

1. Plan for success.
   —Plan the process.
   —Make the process visible.
   —Develop the preplanning matrix.
   —Develop a detailed House of Quality on key opportunities.
2. Improve the process.
   —Develop new processes and alternatives.
   —Execute Pugh Concept Selection.
   —Try out new processes and equipment.
3. Hold the gains.
   —Document the critical process steps.
   —Standardize the process.
   —Publicize the results.

ASI has also adapted the QFD four-phase model for use with continuous batch processes.

QFD AT WORK

The following real-life examples illustrate how novel the application of QFD can be. The company names have been changed in respect of proprietary information.

J&N Corporation

J&N Corporation, a large consumer goods manufacturer, had access to more market research than it knew what do to with. Sifting through this market research and trying to glean something meaningful from it was a laborious and often subjective task. QFD provided a way to organize the data so that J&N could effectively use it.

By using the preplanning matrix shown in Figure 5–3, J&N sharpened its product improvement efforts. It used the customer requirements, an importance rating based on customer input, and a customer competitive assessment.

Although the first customer want (A) was very high in importance, the competitive assessment showed that J&N was trailing the competition in this area. Thus, it would probably want to make an improvement.

The second customer want (B) was also high in importance, but J&N was the leader here, with the key competitors doing the trailing. Hence, the company decided to do nothing except maintain its lead.

The third customer want (C) had middle-of-the road importance. Since J&N was middle-of-the-road in terms of competition, no action was deemed necessary.

The fourth customer want (D), which J&N was also leading, wasn't that important to the customer. Although J&N

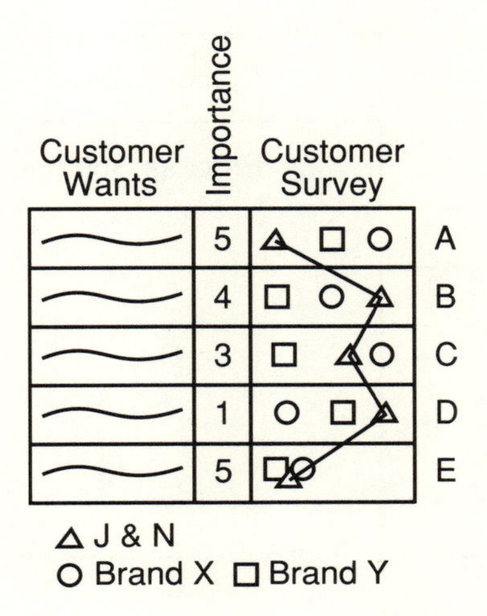

FIGURE 5–3

J&N Corporation sharpened its product improvement efforts via these customer requirements, an importance rating based on customer input, and a customer competitive assessment.

didn't wish to hurt its position here, it was willing to compromise in order to make a strong improvement elsewhere.

The fifth customer want (E) is an important requirement that neither J&N nor its leading competitors had satisfactorily addressed. Thus, this want presented a real opportunity for J&N to leave the competition behind.

B&B Steel

B&B Steel, a major U.S. steel company, wanted to build a new mill. Because steel mills are designed so infrequently and are so expensive to build, not to mention very difficult to change once they have been built, forethought was of utmost importance.

B&B selected QFD as the best tool to solve this problem. Its team created a QFD chart (see Figure 5–4) in which the

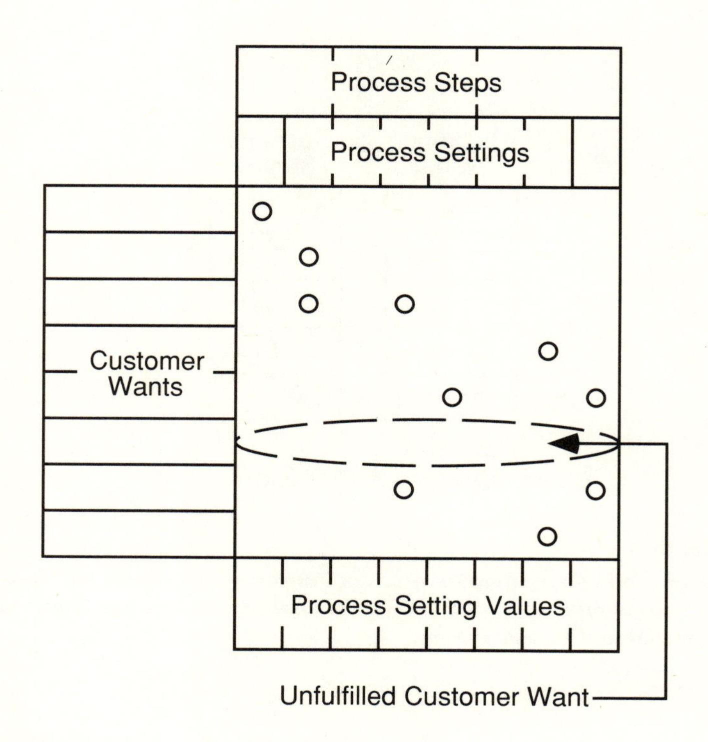

FIGURE 5–4
B&B Steel used QFD to foolproof the design of its new steel mill.

what items were the customer wants for the steel to be produced. These covered a wide range, since steel mills make many types of steel. The *how* items were the process steps and the process settings for each of the process steps, and the *how much* items were the setting values.

After completing the chart, the B&B QFD team noticed some blank rows, which indicated unfulfilled customer wants. This lead to a number of improvements in the company's new mill design, improvements that would have been extremely difficult or downright impossible to achieve after the mill had been built.

Baker Electric Company

The Baker Electric Company, a small but prospering electric power utility, had an unusual problem. Whereas most large companies can wield a big stick over their suppliers, most of Baker Electric's suppliers were larger than Baker Electric itself. The primary supplier that Baker Electric purchased its transformers from, for example, was huge—so huge that it considered Baker Electric "small."

So Baker Electric decided to use QFD for itself, identifying the customer requirements for a substation transformer (see Figure 5-5). These requirements became the *what* items and the actual transformer specifications became the *how* items. Based on this data, Baker

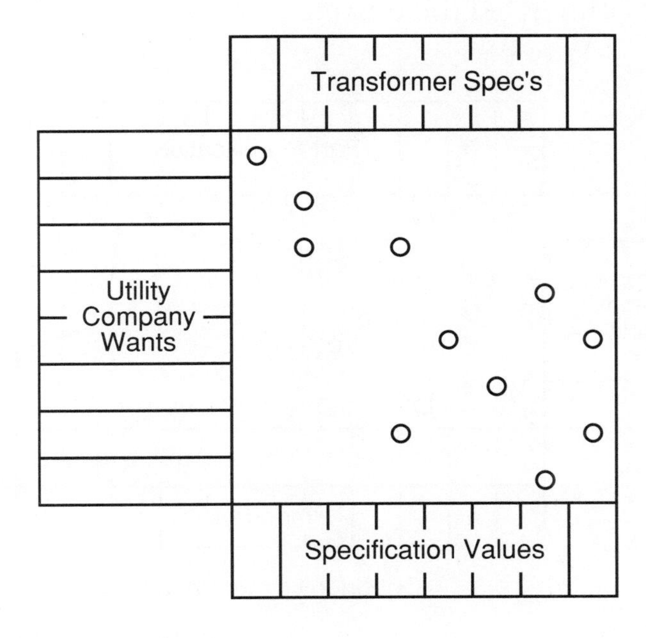

FIGURE 5–5

Baker Electric Company's QFD study yielded a set of specifications that it passed along to its much larger supplier.

Electric put together a comprehensive set of specifications that it then gave to the supplier, thus ensuring that its "small" voice be heard.

Ashton Corporation

Ashton Corporation, a large mechanical device manufacturing company, established a cost-reduction program that included reducing the costs of some of its complex parts. These parts tended to have a multitude of specifications. Because the parts were considered critical, most of the specifications automatically had tight tolerances.

The QFD chart shown in Figure 5–6 provided a closer look at the parts and their specifications. Ashton's QFD team completed a chart for each part with cost-reduction potential and listed the customer wants for that particular

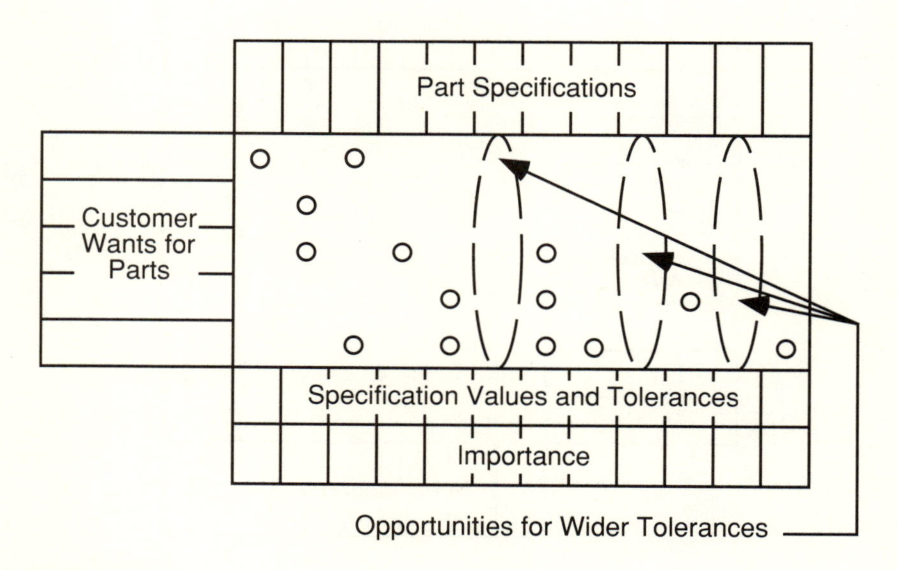

FIGURE 5–6

Ashton Corporation's cost-reduction program benefited from QFD analyses of critical and noncritical parts.

part. These included what the part was supposed to do for the customer, that is, the *what* items. The *how* items were the specifications that came right off the part's blueprint.

The resulting importance ratings for each of the *how* items gave an indication of the relative importance of that item to the customer. In many cases, the columns came up blank—the specifications had nothing to do with satisfying customer wants.

This actually wasn't an error; many of the specifications simply had to be there in order to manufacture the part. Because these specifications weren't crucial to satisfying the customer wants, their tolerances were opened up, which resulted in significant cost savings. Specifications with many symbols, however, were considered critical, and their tolerances were maintained.

Saranen Enterprises

Saranen Enterprises wanted to participate in the European Community market when it opened in 1992 but needed a European partner. Being a small company, and having heard a number of horror stories about joint ventures, Saranen set out to systematically find the right partner.

By using QFD, Saranen Enterprises identified its expectations of a good partner (see Figure 5–7). The *how* items identified if a potential joint venture partner met those expectations. These were specific questions that Saranen Enterprises could ask and verify instead of relying on the spoken word and trust.

Through this process, Saranen Enterprises identified three prospective joint venture partners and then narrowed the three to one. Within two days of that decision, it had negotiated a successful joint venture arrangement.

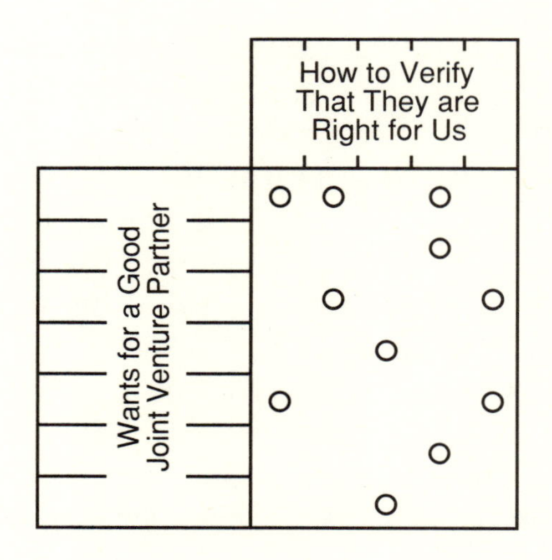

FIGURE 5–7
Saranen Enterprises applied QFD to the identification of a joint venture partner with which it could participate in the European Community market.

Moreover, the joint venture partner was astounded that Saranen Enterprises had asked the right questions, questions that the partner hadn't considered asking.

These examples are based on actual uses of QFD by American companies. A more detailed example of how one leading U.S. utility applied QFD to its response to customer calls process is found at the end of this chapter.

A BLUEPRINT FOR YOUR QFD

A comprehensive description of how to do QFD is beyond the scope of this book. The following eight steps, however, provide a basic road map to QFD implementation. Definitions for some of the technical terms are found in the glossary.

Step 1: Choose the project scope

A core team of QFD participants should be selected to define and deploy the project scope. Many first-time QFD efforts are simply too big and overwhelming. The fundamental assumption behind QFD is an important one. The average company's normal operating system works well enough most of the time or that company wouldn't be in business. A key question to ask is: What do we expect QFD to accomplish that our normal product development process would not? If no good answer exists, the project scope probably isn't correct. Several aspects of the product development process are typically risky and do require special attention from QFD.

Step 2: Identify customers and their wants

Every company has a full series of customers (multiple customers), no matter how simple its product or market. These customers can be viewed as a value chain extending from the company itself to its various distributors and customers, the end users, and, in more and more instances, product disposal. (This is particularly true in Europe, where disposal of excess packaging materials is increasingly difficult, and in the medical products field due to biohazard considerations.)

Customers come in two varieties: those the normal product development process can handle and those with special needs. In addition, some of the latter have high-risk special needs. Once identified, these needs (customer wants) will typically flow into the House of Quality. At this point, the Kano model (see chapter 4) can be used to explore basic performance and excitement features and achieve maximum competitive advantage.

Step 3: Determine the technical requirements

The customer wants identified in step 2 can now be translated into technical requirements, an internal engineering restatement of the wants. These become the specifications for the overall product. The *what* items (customer wants) from the House of Quality will be mapped into the *how* items (technical requirements), relationships will be defined between the *what* and *how* items, and values will be determined (the *how much* items).

Determining the *how much* values (outputs of the QFD chart) often requires additional rooms in the House of Quality. These additions include the rooftop (to look at technical conflicts between specifications), competitive assessment graphs to depict the state of the marketplace, and importance ratings to assist prioritization.

Step 4: Create a design concept

The first phase of the House of Quality should now be complete. The next phase, design deployment, identifies the best design concept, the one that satisfies the technical requirements at the lowest possible cost. Basic creativity fueled by competitive benchmarking and tools like VAVE and Design for Assembly are useful at this point.

A variety of concept alternatives, the seeds of the ultimate design, will result from these efforts. The challenge is to determine which is the best concept. Pugh Concept Selection can be applied here, with technical requirements, cost, and important business issues all influencing the selection. This approach contrasts the possible concepts, resulting in a synthesis of new, improved ideas and, ultimately, the best design.

Step 5: Choose part and material values

The concept can now be detailed out, right down to the bill of materials in which parts and materials are defined. The part characteristics and material qualities that make the product design achieve the technical requirements are placed in the top of the design deployment matrix (the *how* items). A process FMEA can be used to identify which of the many part characteristics are particularly critical. Taguchi Methods can then be applied to help identify the values for the part characteristics (the *how much* items). This activity concludes design deployment, the second QFD phase.

Step 6: Select the process

Step 6 begins the third QFD phase, process planning. Here, the process steps used to create the fabricated parts and assemble them are identified. Existing processes will typically be used so as to avoid the capital investment that accompanies new equipment. Although new equipment should be purchased when capacity or capability dictates, cost implications and process capability must be carefully considered. The individual process steps can then be mapped out.

Step 7: Choose the process parameter values

Next, the critical process parameters for each process step, for example, those knobs and dials that are adjusted to make the process operate effectively, are identified. The parameter values can then be determined. Less critical values will be based on current production practices;

more critical ones determined by Taguchi Methods. This concludes the third QFD phase.

Step 8: Choose the control methods

The last step concerns the fourth QFD phase—controlling manufacturing operations. Types of human intervention to help control the process on a daily basis can now be identified, just as the settings for the process parameters were determined in step 7. Step 8 addresses practices and procedures rather than engineering issues. These are based on a risk assessment driven by a process FMEA or, if not available, the QFD team's best judgment. The control methods will fall into four categories: quality assurance, preventive maintenance, mistake proofing, and standard operating practices that ensure consistency in the operation.

These eight macro steps define a road map for QFD implementation and a means of tracing critical part characteristics through the manufacturing process. Organizations in the aerospace industry that must comply with Boeing's D1-9000 requirement will find a natural means of ensuring traceability in QFD. Organizations in other industries will similarly benefit.

QFD at FPL: At Your Service

"Quality begins and ends with meeting the needs of the customer," wrote the late Duane E. Orlowski (1990). "QFD's potential for improving customer satisfaction in service industries is enormous and as yet virtually untapped."

Florida Power & Light Company (FPL), the only American winner of the Deming Prize, has been tapping into the power of QFD for some time now. FPL, a division of the FPL Group, has applied QFD to the increasingly competitive public communications industry by deploying customer requirements (*what* items) and quality characteristics (*how* items) into job functions throughout its service organization.

Consultants from Qualtec of Juno Beach, Florida, also a division of the FPL Group, assisted with the implementation of QFD at FPL. Qualtec consultants J. L. Webb and William C. Hayes (1989) examine two of FPL's QFD applications: (1) deploying corporate quality elements via a "Customer Needs Table of Tables" and (2) revising the "response-to-customer calls" process at regional phone centers. The following example is primarily based on the second application.

Background

Customers in the area covered by the Southern Division Regional Phone Center, Miami, Florida, generated approximately 40,000 calls per week. FPL employees typically answered five types of calls: requests for payment extensions, requests for new services, questions about billing statements, reports of power outages, and complaints relating to any of the above. Additionally, FPL employees had to be able to respond to such calls in three different languages.

Prior to the use of QFD, the following process was in effect:

1. A customer called FPL for assistance.
2. An FPL employee took the call, responded to the customer request, and performed any necessary follow-up.

FPL had previously used a quarterly customer service satisfaction survey to assess customer satisfaction. Problems had been noted in three categories: (1) courteous and professional manner, (2) caring and concern, and (3) ability to answer questions. The Southern Division Regional Phone Center had scored low in all three categories.

Creating Matrices

FPL's QFD team established primary, secondary, and tertiary customer requirements (that is, increasingly specific *what* items) related to the response-to-customer calls process. The primary and secondary *what* items were drawn from an earlier QFD study; the tertiary items from verbatim responses to open-ended questions on the customer service satisfaction survey. These items were then grouped using the affinity process, one of the Seven Management Tools (see glossary).

The primary *what* items were "considerate customer service" and "accurate answers/timely actions." The secondary *what* items for "considerate customer service," for example, were "fair treatment," "courteous, friendly employees," and "concern for customer problems." These items were then further subdivided into tertiary *what* items. For example, "courteous, friendly employees" was broken down into "understandable speech," "professional manners," and "customer courtesy."

The team created the matrix shown in Figure 5–8 in order to examine the relationship between the customer requirements (*what* items) and the quality characteristics (*how* items). The customer rating indicated the need to improve the time required to answer phone calls. An action plan and countermeasures were implemented that resulted in a fourfold improvement on performance.

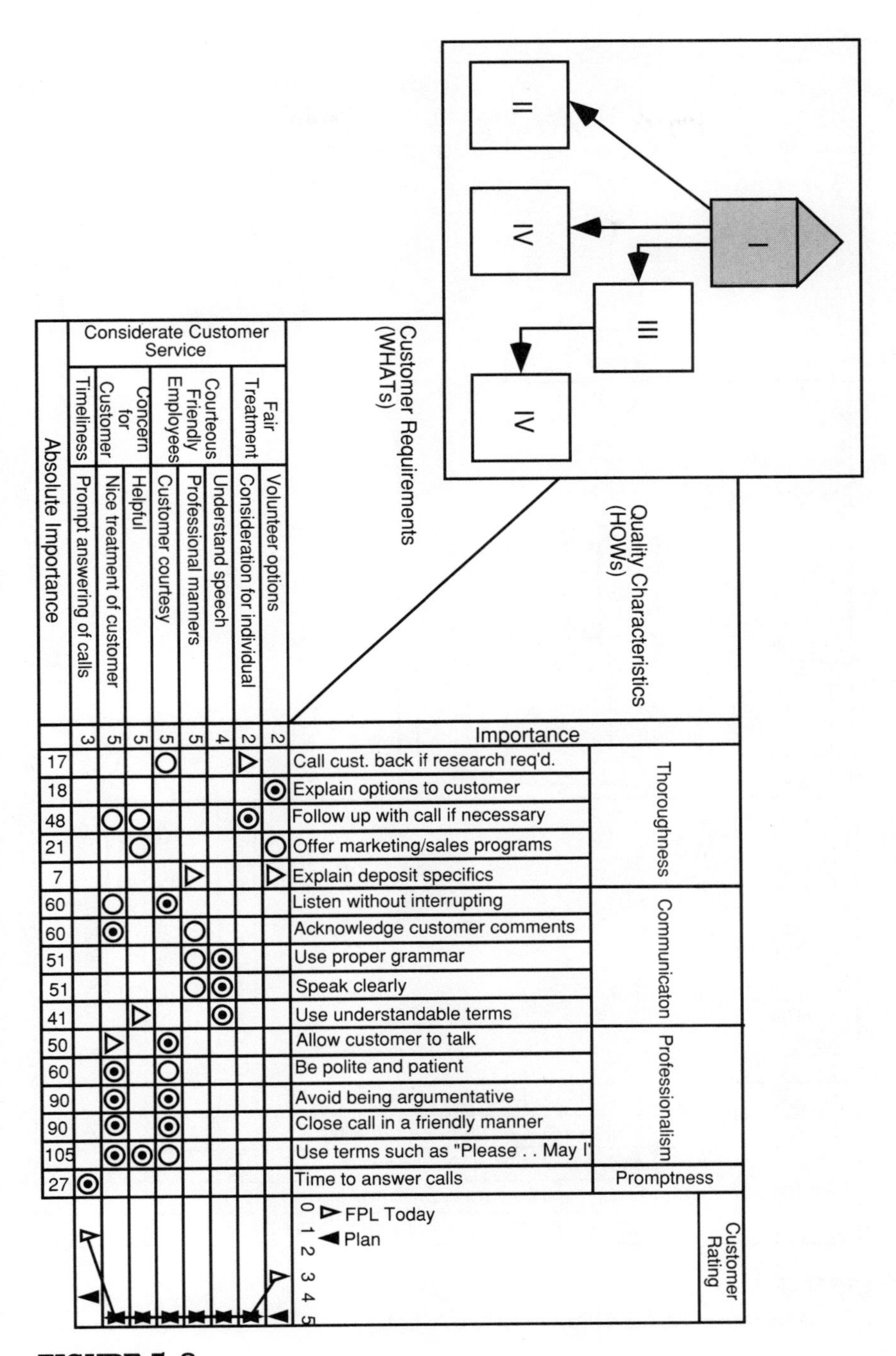

FIGURE 5–8

FPL—Service planning example.

Next, the team analyzed the key quality characteristics in relation to FPL's existing training program and employee selection process and created related training and hiring matrices (see Figures 5–9 and 5–10). Upon doing so, the team concluded that the key quality characteristics were not being taught in the training program or being considered during the hiring of new employees.

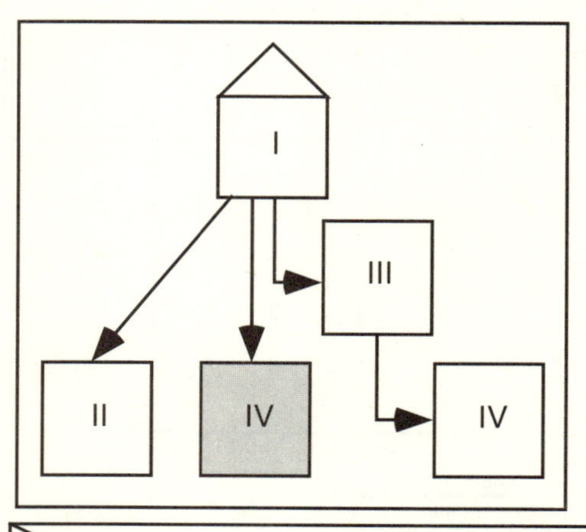

Quality Characteristics (WHATs) \ Training Modules (HOWs)		Customer Courtesy Pays	FPL Organization	Billing Systems	Utility Industry Overview	Procedure Manual	Collections
Thoroughness	Explain options to customer						○
	Follow up with call if necessary					△	
Professionalism	Avoid being argumentative	⊙					
	Close call in a friendly manner	○					
	Use terms such as "Please . . May I"						

FIGURE 5–9

FPL—Education and training matrix examples.

The Revised Process

As a result of the QFD process, FPL made the following changes.

1. The service observation form used to monitor customer service calls was revised so as to link the customer requirements with employee performance, providing

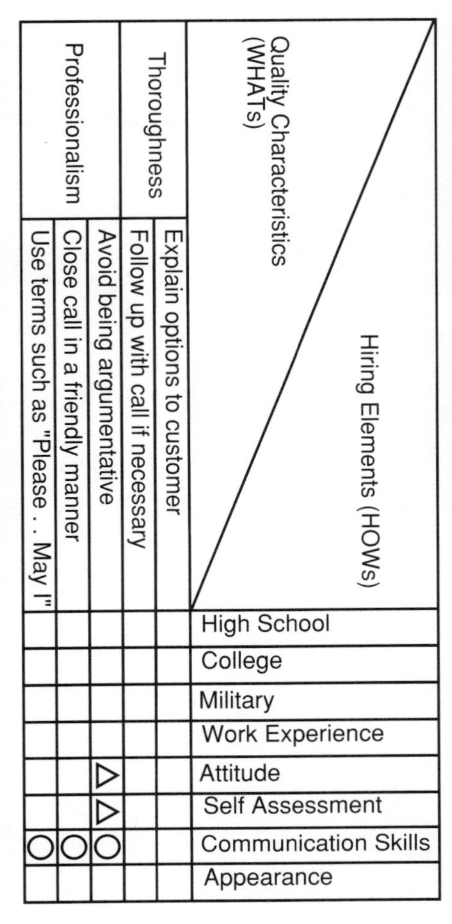

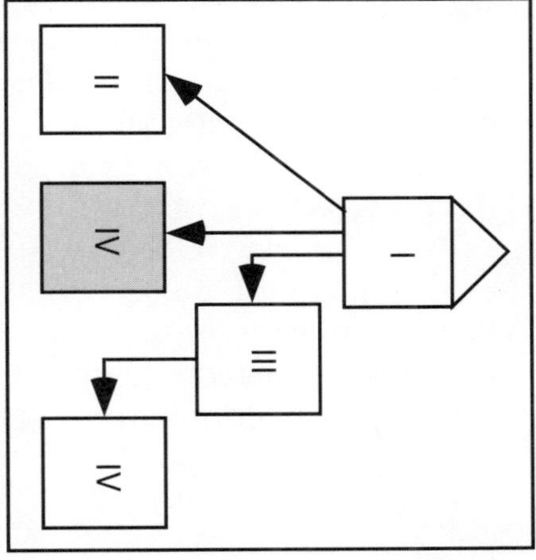

FIGURE 5–10
FPL—Hiring matrix example.

objective evaluation of responses to customers and timely and valid feedback to FPL employees.

2. The key quality characteristics were incorporated into the existing training program, and customer requirements and job requirements were clearly defined.

3. An interview analysis check sheet was created for use when hiring new employees. The check sheet is used to evaluate each interviewee's ability to perform the required job. The interview itself included role playing, which helps the interviewers assess the interviewee's personal skills and reactions and gives the interviewee an idea of the job requirements.

Since the application of QFD to FPL's response-to-customer calls process, customer satisfaction has improved dramatically, and customer complaints to the Florida Public Service Commission have been significantly reduced. Additionally, regional phone center management has become more aware of how different activities affect customer satisfaction, specific problems have been passed on to quality improvement teams, and additional QFD studies have been initiated in service, software, and product development areas.

Excerpted from "QFD in the Service Environment," by Kurt R. Hofmeister, *ASI Journal*, Fall 1990, pp. 19–34.

REFERENCES

Ehrlich, D.M., and D.J. Hertz. 1993. *Applying QFD to Health Care Services: A Case Study at the University of Michigan Medical Center.* Paper read at the Fifth Symposium on Quality Function Deployment, 21–22 June, at Novi, Michigan.

Eureka, W.E., and N.E. Ryan. 1990. *The Process-Driven Business: Managerial Perspectives on Policy Management.* Dearborn, Mich.: ASI Press.

Hofmeister, K.R. 1990. "QFD in the Service Environment." *The ASI Journal.* (Fall): 19–34.

Irving, K. 1993. Telephone interview. Interviewed by Nancy Ryan.

Orlowski, D.E. 1990. Preface to *The ASI Journal.* (Fall): 3–4.

Webb, J.L., and W.C. Hayes. 1990. *Quality Function Deployment at FPL.* Paper read at the Second Symposium on Quality Function Deployment, June 18–19, at Novi, Michigan.

Chapter Six

The Power of QFD

Quality Function Deployment derives much of its strength from its ability to quench an unpopular phenomenon—Murphy's Law. Quite simply, QFD helps keep things from going wrong as a product makes its way through a complicated series of design and production activities.

QFD also addresses the adage "Things that go without saying usually go without doing." By creating a disciplined outline for all involved to follow, QFD enforces efficient information sharing. This, in turn, enhances the product development process.

A MATTER OF TIME

QFD can help improve quality while decreasing costs and, perhaps more importantly, by reducing time to market. Cost and quality improvements have become givens in today's competitive market, making timeliness that much more important.

By using QFD and Taguchi Methods, many industry leaders have improved their business operations while also improving quality and achieving remarkable time savings. Taguchi Methods are combined engineering and statistical methods designed to achieve rapid improvements in cost and quality by optimizing product design and manufacturing processes. See the appendix for more details.

Through the use of QFD, the percentage of quality defects attributed to product design can be substantially reduced without the need for inspection, which is an expensive, ineffective form of quality control that is both outmoded and counterproductive. QFD can also be used to initiate a healthy dialogue between purchasing and suppliers aimed at reaching the optimum in product or material cost and quality.

QFD represents a change from the inspection-rejection-rework-scrap form of quality control to one that begins at the onset of the product development process. QFD replaces the reactive, firefighting approach to quality control with one that is proactive and preventive in nature. In other words, quality doesn't have to equate to high cost. Quality does cost more if it's achieved via inspection. When designed into a product, however, quality actually reduces costs.

This upstream effort does away with the inefficient design-test-fix scenario, the design counterpart of product inspection. When testing reveals a major flaw in a prototype design, time and dollars are wasted. When this scenario repeats itself two or three times—the dreaded oh-not-again syndrome—the expenditure is doubled or even tripled. An adequate design will eventually result, but at great expense. When products are designed, not fixed, to meet customer requirements, both costs and product development cycles are reduced. Inspection to achieve quality is no longer relevant during product design or manufacture.

All product development endeavors are investments. But when a company enters the reactive find-and-fix mode, last-minute changes increase the investment and delay the desired payback. With QFD's proactive approach, such problems are avoided, and product introduction and payback occur that much sooner (see Figure 6–1).

According to MIT's Don Clausing (1988), QFD addresses three broad problems attacking American industry: (1) disregard for the voice of the customer, (2) loss of information as a product moves through the development cycle, and (3) different interpretations of specifications by the various departments involved. QFD also provides solutions to two related problems: division by department and sequential timing.

The negative effects of division by department are diminished as QFD is implemented cross-functionally. Members of the QFD project team work together, not as separate entities. Additionally, QFD's vertical deployment ensures that a concurrent, not sequential, approach to product development occurs.

One of QFD's most-cited benefits is its ability to generate team involvement that's sustained over the entire product development cycle. The results of this team synergy are much greater than the sum of the team's parts. QFD is a systematic way of bringing the collective wisdom of the corporation to bear on a problem. By pooling the knowledge of the QFD project team, enhanced decision making occurs; personal biases disappear as the team begins functioning at top capacity.

"The strength of QFD is that the process itself becomes a catalyst that generates team effort and cooperation," said Calvin W. Gray (1988), president of the Handy & Harmon Automotive Group in Auburn Hills, Michigan. "QFD then becomes a mechanism for communication among the various areas working on the project."

This team approach is fitting with American industry's renewed focus on teams and team building. In *Reviving the American Dream* (1992), for example, author Alice M.

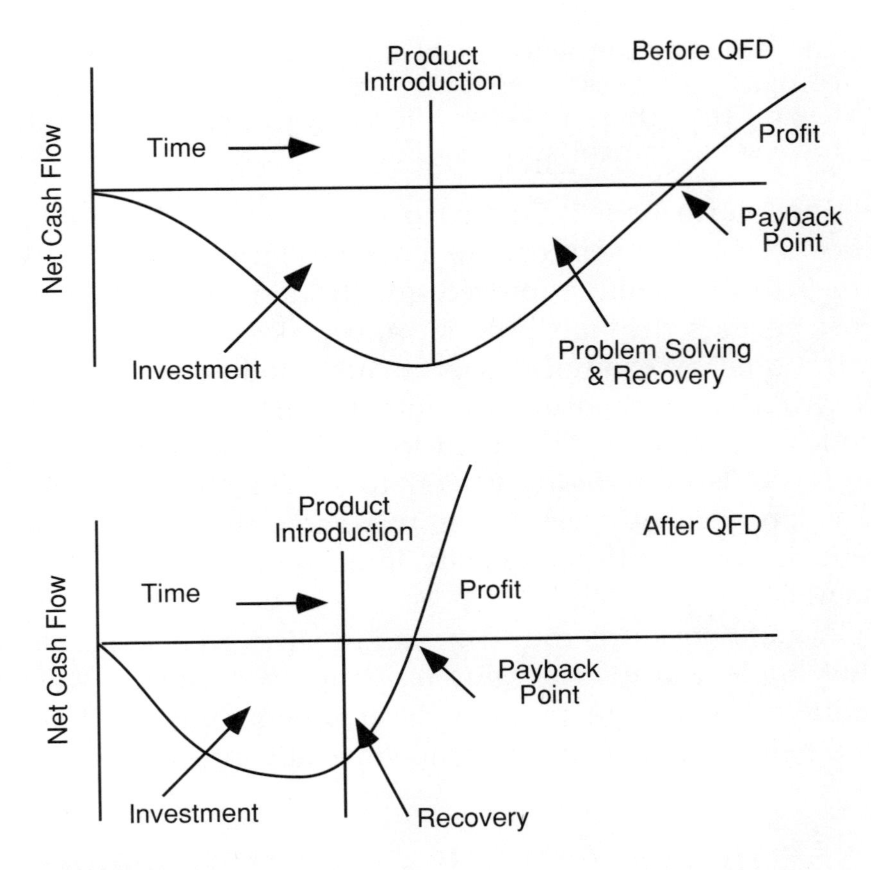

FIGURE 6–1
With QFD's proactive approach, product introduction and payback occur much sooner.

Rivlin notes that many American businesses are restructuring themselves from the ground up in response to forecasts of doom and nonstop foreign competitiveness. Rivlin notes that

> This revolution has put enormous emphasis on improving the quality of products and services, increasing responsiveness to customers and clients,

> and empowering workers at all levels to contribute to company success. . . . Productivity depends on workers feeling that they make a difference to team effort and have some control over results.

Short-term benefits afforded by QFD include shorter product development cycles, fewer design changes, fewer start-up problems, improved quality and reliability, and cost savings through product and process design optimization. For example, Toyota Auto Body Company in Kariya, Japan, reported a cumulative 61 percent reduction in start-up costs related to the introduction of four van models from January 1977 to April 1984. The product development cycle during this same time frame was reduced by one-third. Quality improvements were cited as well.

In addition, numerous Toyota suppliers have reported that implementation of QFD has improved quality while reducing costs, cut product development times in half, and helped achieve major competitive advantages.

THE VOICE OF THE CUSTOMER SPEAKS

According to Rivlin,

> Big American companies pioneered mass production and assembly line methods in the 1920s. Now the emphasis has shifted to more decentralized management, greater emphasis on teamwork, worker creativity, and interaction with customers to ensure high quality and satisfaction.

Customer requirements can be easily misinterpreted during the complicated product development cycle. For example, the marketing manager may ask for "understated opulence." To the design engineer, however, understated, or, for that matter, opulence, may have a completely different meaning. The QFD process translates the customer's message in its purest form, without the ambiguity often caused by multiple interpretations.

QFD not only focuses companywide attention on customer requirements, it also provides a mechanism to target selected areas where competitive advantages could help improve market share—areas with undeveloped potential. By identifying critical design and part characteristics—the ones that have the greatest influence on overall customer requirements—QFD ensures that product development efforts yield the most bang for the buck.

QFD can also be used to lower warranty costs on existing products, which has a positive effect on profitability. One of the most circulated QFD applications, the Toyota Auto Body rust case study, was initiated in response to escalating warranty costs due to corrosion. Though the use of QFD and Taguchi Methods, Toyota virtually eliminated its corrosion warranty expense.

At Chrysler Corporation, which began using the process in 1987, elements of QFD are being used in conjunction with preplanning for new product platforms and as part of the guidelines for product assurance teams. To paraphrase Chrysler's Robert J. Dika (1993), what Chrysler has learned over a six-year period of studying and applying QFD has now been integrated into the overall process of quality planning.

According to Dika, Chrysler benefited from using the preplanning matrix to establish overall priorities for large projects such as total vehicles and to help break down the task. In addition, teams working on product systems soon learned to appreciate Pugh Concept Selection and FMEA.

Dika emphasizes that diligent small teams can make a difference when it comes to QFD. During the seven-year span described in Dika's paper, "QFD has grown in number of projects and significance, contributing to both the company's fortunes and the development of a more customer-driven, team-based culture."

Although an estimated 51 percent of Chrysler's application of QFD revolves around the House of Quality, a variety of other QFD avenues have also been explored. These include preplanning, concept selection, design deployment, process deployment, and production deployment. And over the six-year period since Chrysler started using QFD, a total of 95 QFD projects have been facilitated.

CREATING A KNOWLEDGE BASE

Effective QFD applications go hand in hand with ongoing continuous improvement. Continuous improvement via QFD occurs when the QFD documents are revisited and information is added that will enhance the next design iteration, thus reducing product development time.

After the efforts of a QFD project have been deemed successful, the knowledge base created for that project serves as a repository for engineering knowledge. QFD teams at Chrysler, for example, used the preplanning matrix developed for the P-body (Dodge Shadow and Plymouth Sundance) for the new model programs that followed. In addition, many of the QFD projects initiated for the PL-body (Dodge and Plymouth Neon) drew heavily from projects previously completed by LH-body (Chrysler Concorde, Dodge Intrepid, and Eagle Vision) QFD teams.

With QFD, knowledge can be preserved in one place, in contrast to unruly design standards manuals. The matrices and charts prepared during the QFD procedure create a working document that can be easily referenced and learned from.

The QFD knowledge base holds great promise for future product development efforts. It holds the answers to questions concerning what decisions were made and why—and can simplify similar decision-making endeavors. Concurrently, it helps prevent problems that occurred in 1993 from recurring in 1998—and beyond.

In addition, QFD can be used to train entry-level engineers. By reviewing the results of successful QFD projects, entry-level engineers begin higher up on the learning curve. The process also offers potential for continuing education of employees across departmental lines.

According to Handy & Harmon's Calvin W. Gray, QFD provides a database for design engineering of future products. Gray noted that while QFD may initially consume more man-hours, from an engineering-talent standpoint, the hours required for the same level of product quality on similar, subsequent products will be much less. The net result? A significant reduction of time from product conception to introduction in the field.

The Budd Company's Norman E. Morrell (1988) agrees:

> QFD lays a foundation for future work—you don't have to continually reinvent the wheel or wonder how you did it last time. It also provides a needed discipline, much like a pilot's checklist. A guy may have flown 300 missions, but he still checks off every instrument when he gets into that cockpit.

Because QFD unlocks what Mazda calls the "hidden knowledge" of the organization—that is, knowledge that is locked deep in the mind of the employees—the organization becomes a learning organization instead of an organization of learned people. At leading companies worldwide, QFD is used to make good engineers great engineers.

INTEGRATING PROCESS AND DESIGN

QFD can be thought of as the glue that binds the various product development stages. By tying design and process activities together, it provides integration of the various functions.

Raymond P. Smock (1988), retired manager of advanced quality concepts development and product assurance at Ford Motor Company, described QFD as an all-encompassing planning framework for product development. "QFD integrates the process by which you translate customer requirements into technical requirements for each stage of product development," he explained. "QFD prioritizes product and manufacturing process characteristics and highlights areas requiring further analysis." In Smock's opinion, QFD is of greatest benefit when applied to complex systems that don't lend themselves to such traditional approaches as rule-based design.

QFD is being taught at Ford in conjunction with a Total Quality Management (TQM) training program begun in June 1987. The groundwork for this program—which is the result of extensive study and evaluation—began in the early 1980s, when the late quality guru Dr. W. Edwards Deming visited Ford, followed by the late Japanese TQM authority, Dr. Kaoru Ishikawa.

QFD is highly complementary to TQM and Companywide Quality Control (CWQC) programs, as well as simultaneous engineering partnerships. (Depending on the industry, simultaneous engineering also goes by concurrent engineering, integrated product development, or other company-specific names.)

According to the *Christian Science Monitor* article "What Is Total Quality Management," 3 May 1993, Louis Lataif, dean of Boston University's School of Management and a former Ford vice president, has coined "five facets of TQM." They are: (1) customer focus, (2) management by facts, (3) continuous improvement, (4) total investment, and (5) systematic support.

QFD has the potential to address each of these five facets. It is also a template for implementing simultaneous engineering, providing a framework for application of the manifold tools and activities required to develop products in a collaborative environment. It promotes the same up-front team approach to concurrent product development while placing increased emphasis on the voice of the customer and providing detailed documentation of the team's efforts.

According to MIT's Don Clausing, who promotes the use of QFD with concurrent engineering, the cultural change required for effective cross-functional communication has not yet happened in many organizations. Thus, QFD has the potential to significantly enhance the concurrent engineering strategy. QFD's multifunctional team approach supports and connects many decision-making processes throughout the development project.

How important is QFD to a total quality program? When asked what elements led to the quality transformation at Toyota Auto Body, Akashi Fukuhara (1988) named Taguchi Methods, FTA and RFTA, FMEA, SPC, and QFD. Fukuhara, director of the Central Japan Quality Control Association (CJQCA), worked at Toyota Auto Body for some 20 years and was instrumental in instituting Total Quality Control at the company.

Fukuhara was also asked to rate the importance of each of these tools to quality improvement at Toyota Auto Body. He made the following estimates: Taguchi Methods, 50 percent; FTA/RFTA, 35 percent; and FMEA,

15 percent—which total 100 percent! SPC, he explained was used to monitor, maintain, and uplift quality, rather than for actual quality improvement. This doesn't mean that SPC should not be used for quality improvement in America, but that Toyota's processes no longer require the use of SPC for quality improvement. QFD, on the other hand, was the map that identified when and where to apply the above tools (see Figure 6–2).

This is because QFD not only helps identify which analytical tools would best be utilized during the product development cycle, but it also identifies conflicting company measures that would benefit from optimization. This feed-forward mechanism captures knowledge that might otherwise go undocumented (see Figure 6–3).

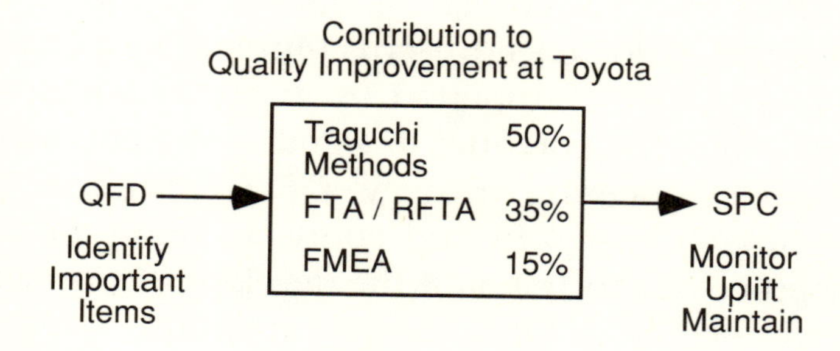

FIGURE 6–2

Contributing to quality improvement at Toyota Auto Body were Taguchi Methods™ (50%), FTA/RFTA (35%), and FMEA (15%). SPC is used to maintain quality; QFD to identify where to use Taguchi Methods, FTA/RFTA, and FMEA.

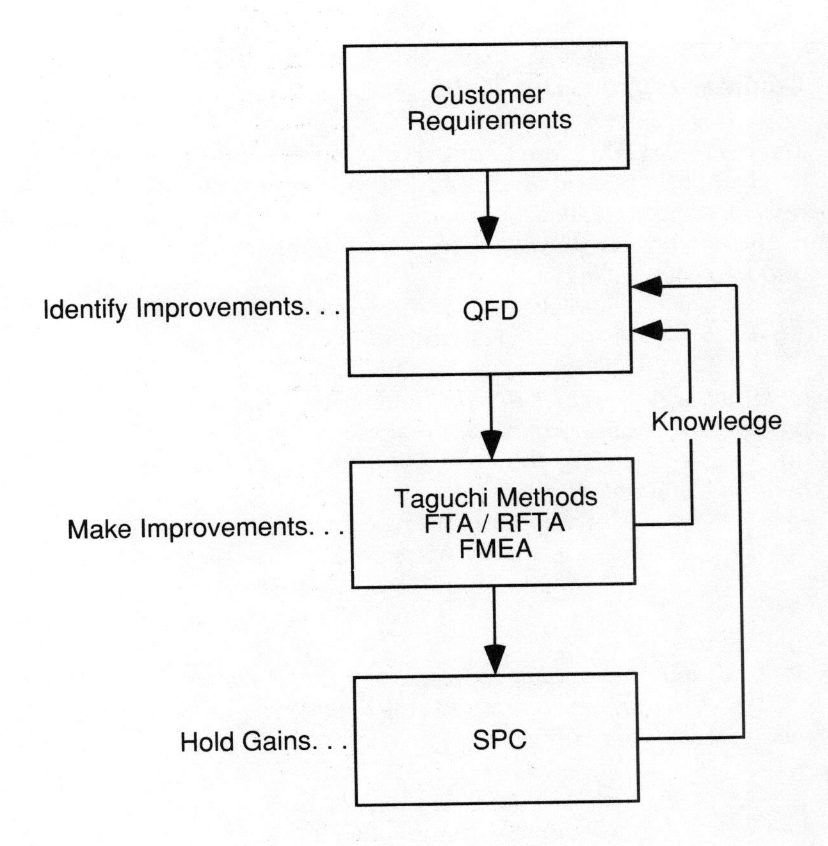

FIGURE 6–3

QFD employs a feed-forward mechanism that identifies where to use such tools as Taguchi Methods™, FTA/RFTA, FMEA, and SPC and then documents their use.

Comments from the Field

The need for QFD is best summed up by the statement "get better or get beat." QFD will simply improve and focus new product development activities by getting the whole organization to concentrate on rapid development of new products with high quality and low cost.

—Michael E. Chupa, vice president of marketing,
ITT Automotive, Auburn Hills, Michigan

QFD helps tie product design and process activities together—it provides better integration of the various functions. It's a systematic way of bringing the collective wisdom of the corporation to bear on the problem.

—Dr. Don Clausing, Bernard Gordon Adjunct Professor
of Engineering Innovation and Practice,
Massachusetts Institute of Technology,
Cambridge, Massachusetts

We can't improve quality by continuing to reactively fix problems. QFD gives us an opportunity to stop talking about fixing and start talking about preventing.

—Robert J. Dika, quality planning executive—
small car platform, customer satisfaction
and vehicle quality office at Chrysler Corporation,
Auburn Hills, Michigan

QFD formalizes your product development process and the recordkeeping that goes along with it. This helps everybody understand just what the process entails.

—James T. Gipprich, sales manager
of the light duty vehicle brake group,
Kelsey–Hayes Company, Romulus, Michigan

What's the single greatest strength the Japanese have gained from using QFD? Ensuring that the product they deliver to the market comes the closest to fulfilling all needs the customer might experience.

—Calvin W. Gray, president,
Handy & Harmon Automotive Group,
Auburn Hills, Michigan

As a professor, I'm always looking for concepts that are teachable. QFD is a teachable discipline. One can teach it in a classroom or elsewhere, providing people with the basis for getting started or changing the way they do these activities.

—Walton M. Hancock, William Clay Ford Professor
of Product Manufacturing, University of Michigan,
Ann Arbor, Michigan

QFD is an excellent planning tool—a vehicle by which we can get people working together as a team to reach common goals. It's a tool that ensures that all bases are covered, that no stone is left unturned.

—Norman E. Morrell, corporate manager
of quality-product reliability,
The Budd Company, Troy, Michigan

QFD offers many benefits for the ultimate consumer because of what it accomplishes up front. With QFD, there's no need to put on bandages. Doing so usually costs everybody money—despite all efforts to ensure that those bandages solve the problem, weaknesses generally exist. The ultimate customer suffers in the long run, as does the company's reputation.

—George R. Perry, vice president and general manager,
Siemens Automotive, Newport News, Virginia

QFD forces people to be more conscious of potential problems and, as a result, to do a better job. Not only will QFD help us find the holes, it will help us from repeating past mistakes.

—Robert M. Schaefer, director of systems engineering methods, General Motors Corporation, Warren, Michigan

QFD will point you toward the areas where you need to do further analytical work—it results in more focused analyses aimed at meeting customer requirements. This is a major breakthrough. People are searching for where to use analytical tools, such as design of experiments. QFD leads the way.

—Raymond P. Smock, retired manager of advanced quality concepts development and product assurance, Ford North American Automotive Operations, Dearborn, Michigan

One of QFD's biggest merits is in supplying new employees with the background and experience gained by their predecessors. Internally, we see a smoother transition through QFD and, therefore, reduced lead times for new designs. In addition, the products we manufacture will be of higher quality.

—Peter J. Soltis, retired senior technical specialist of product engineering, Kelsey–Hayes Company, Romulus, Michigan

REFERENCES

Chupa, M.E. 1988. Personal interview. Interviewed by Nancy Ryan.

Clausing, D.P. 1988. Personal interview. Interviewed by Nancy Ryan.

Dika, R.J. 1993. *QFD Implementation at Chrysler: The First Seven Years.* Paper read at the Fifth Symposium on Quality Function Deployment, 21-22 June, at Novi, Michigan.

Dika, R.J. 1988. Personal interview. Interviewed by Nancy Ryan.

Fukuhara, A. 1988. Personal interview. Interviewed by Nancy Ryan.

Gipprich, J.T. 1988. Personal interview. Interviewed by Nancy Ryan.

Gray, C.W. 1988. Personal interview. Interviewed by Nancy Ryan.

Hancock, W.M. 1988. Personal interview. Interviewed by Nancy Ryan.

Morrell, N.E. 1988. Personal interview. Interviewed by Nancy Ryan.

Perry, G.R. 1988. Personal interview. Interviewed by Nancy Ryan.

Quality Function Deployment for Hardware Implementation Manual. 1992. Dearborn, Mich.: ASI Press.

Rivlin, A.M. 1992. *Reviving the American Dream.* Washington, D.C.: The Brookings Institute.

Schaefer, R.M. 1988. Personal interview. Interviewed by Nancy Ryan.

Soltis, P.J. 1988. Personal inteview. Interviewed by Nancy Ryan.

Smock, R.P. 1988. Personal interview. Interviewed by Nancy Ryan.

ter Haar, S., D.P. Clausing, and S.D. Eppinger. 1993. *Intregation of Quality Function Deployment and the Design Structure Matrix.* Massachusetts Institute of Technology.

Chapter Seven

Getting Started

When the first edition of this book came out in 1988, Quality Function Deployment was in its infancy in the United States. It is now a toddler. The majority of the QFD pioneers interviewed for the first edition said that QFD was in the learning-curve stage at their respective companies. In many ways, it still is. These companies have come to realize two things: (1) it takes a good number of years to implement QFD across the board; and (2) long-term quality improvement is never easy.

Most of the QFD pioneers learned of the process through seminars or study missions to Japan. All expressed enthusiasm for the process. All cautioned that putting QFD to work in the United States wouldn't be easy, and that doing so would require demonstrated organizational change—which still rings true today.

Handy & Harmon's Calvin W. Gray (1988) explained:

> Many companies think they're ready for this type of change. But when it comes time to practice it, they don't really sign up. Everybody is for improvement—but they're not always ready to make the changes that must precede improvement. It's important for management to understand that change may be required to successfully implement QFD.

FIRST STEPS

Chrysler's Robert J. Dika uses higher education jargon to describe five levels of QFD, each of which contributes significantly more to the product development process. At the first level, the "100 level," people begin to informally think about customer requirements and respond to them cross-functionally. The "200 level" brings about more qualitative market research and team formation. The "300 level" moves toward the quantitative, with formal meetings and teamwork leading to the construction of well-planned QFD matrices.

In *QFD Implementation at Chrysler: The First Seven Years*, Dika (1993) described the "400 level" as a "technological weapon to fight for market share. . . . QFD is used to identify critical areas and other quality and reliability tools are used to solve the problems identified by the QFD projects." QFD is used to facilitate creative learning at the post-graduate "500 level," with teams expanding the QFD envelope by discovering new ways to use it.

Regardless of the level at which it is introduced, training in QFD should be targeted to fit the needs of the organization. For instance, a modular approach works best for GE Aircraft Engines in Cincinnati, Ohio. Instead of teaching QFD in an intensive three-day workshop, this commercial and military aircraft manufacturer initially introduced QFD via half-day overviews. The participants then returned to their respective areas and began applying QFD to specific problem areas.

Two months later, the participants regrouped for a brainstorming session. This approach led to the creation of specific training designed to meet specific problems in a flexible, modular format. The resulting training program, designed by a consultant from the American

Supplier Institute (ASI), consisted of four half-day modules: introduction to QFD, the House of Quality, product planning, and production planning. While the introductory module can be termed traditional "training," the other modules are "facilitated work sessions" for actual team projects. According to GE Aircraft Engine's Ralph A. Kirkpatrick (1993), program manager of CS6-80C2 total product quality, the modular approach helps promote acceptance and application by providing just enough training just in time.

A TEAM EFFORT

QFD is designed to be a team activity, from the initial identification of customer requirements through the deployment of those requirements. As a rule, multidisciplinary teams should consist of approximately five to seven people, with all key functions represented. The project leader should be skilled in coordination, not domination, because QFD is consensus oriented and excels in a freewheeling environment.

The nature of QFD collaboration will depend on the nature of each respective team. For colocated teams, the QFD work will blend with the normal work, and QFD meetings will be more ad hoc than formal. For teams with local (but not colocated) members, weekly team meetings will focus on QFD coordination with some QFD work added in. For teams with out-of-town members, formal meetings will be less frequent due to travel constraints and supplemented with subteam activities, conference calls, and the like.

Major projects will probably require four to six hours of meeting time per week to coordinate activities and update matrices and charts. Team members should also devote a good deal of time apart from meetings to work on individual assignments. But much of what they'll do will be part of their regular job assignments, directed by the QFD planning effort, of course.

In general, QFD is usually selectively applied to components or systems that call for a marketing or competitive advantage—high-risk issues. In the majority of instances, if the everyday operating system works, don't try to fix it. In the minority of instances that cry out for QFD, its implementation should be coordinated with the program timing. QFD projects can take from months to years to complete, depending on the projected duration of the product development effort.

There are several key points the QFD project team should remember. The process may look easy, but it requires effort. Many of the entries may look obvious *after* they are written down. The charts may appear to be the objective, but they're really the means of achieving the objective. And the objective, identifying and deploying the voice of the customer, may appear to be deceptively simple.

Selecting a project that's manageable and supported by management and peers will also help get QFD off to a good start. It is best to begin with a project that's small enough to follow through to completion. Success on a small scale always beats failure on a large scale.

Whether acquired via field work or other forms of marketing analysis, true customer requirements—requirements that are clearly expressed in layperson's terms—must be acquired before QFD can begin. Without a true understanding of the voice of the customer, QFD can become a futile exercise. Achieving an understanding of the voice of the customer, however, isn't always as easy as it sounds. But once achieved, it sets the stage for a successful QFD application.

Kelsey–Hayes Company in Romulus, Michigan, answered ASI's call for QFD case studies in late 1985. Kelsey–Hayes' first QFD application, a House of Quality for a coolant-level sensor being developed for Ford Motor Company (see chapter 4), was critiqued during the 1988 ASI study mission to Japan. That led to subsequent QFD charts and a second QFD case study. The second QFD application, a power door lock actuator, entailed extensive brainstorming and review of marketing data.

According to James T. Gipprich, sales manager of the light duty vehicle brake group,

> We were basically looking at two customers, Ford Motor Company and the person who drives the car. We really wanted to identify the needs of the person who uses the total system—not only our component, which is installed within the door. Our team started by identifying what we thought were the customer requirements. Next, we put those requirements in a questionnaire and had people comment on them. That's when we found out that what we thought were appropriate questions didn't make a lot of sense to the average person driving the car.

Once the questions were reworded, the joint Kelsey–Hayes/Ford project got well under way. It was a concerted team effort that include two-hour biweekly meetings with well-planned agendas.

Robert M. Schaefer (1988), director of systems engineering methods at General Motors in Warren, Michigan, also stressed the importance of a crystal-clear interpretation.

> The secret is to capture the voice of the customer in such a manner that the engineer knows what to do with it and unleashes creative energies in answer to

> customer needs. The engineer doesn't really need that interpretation from marketing—it's far better that he or she hears what the customer is saying. That's step one. If you haven't done that step properly, the rest of your QFD project will be questionable.

As explained in chapter 4, a series of steps follows determination of the customer requirements. These include: determining the company measures, relating the customer requirements and company measures, evaluating the competition, prioritizing efforts, identifying and resolving trade-offs, determining values of the company measures, and completing the subsequent QFD matrices and charts.

These matrices and charts can be completed by hand or by computer. Kelsey–Hayes chose the latter.

Peter J. Soltis (1988), retired senior technical specialist of product engineering, at Kelsey–Hayes explained.

> We took the time to format a House of Quality chart on our CAD/CAM system. It took awhile to develop that chart, but it helped all subsequent charts. We can now reduce our preparation time substantially. By using the CAD/CAM system, we can vary the size of the charts depending on the scope of the QFD study.

Such QFD software programs as QFD/Capture (International TechneGroup, 1993) and QFD Designer (American Supplier Institute and Qualisoft, 1991) offer similar options.

IMPLEMENTATION ISSUES

A number of challenges await QFD champions eager to introduce the process to their companies. These include information overload; time, patience, and discipline issues; QFD's Japanese heritage; and the understandable reluctance of other companies to share QFD charts containing proprietary information. In addition, QFD is far from the first product improvement program American industry has been introduced to, and a prove-it attitude often accompanies the implementation of such programs.

Finding the time to complete a QFD case study may also prove challenging. QFD may be perceived as an added workload instead of a better way of doing things. It may get lost in the everyday shuffle; it may be considered too time consuming. Hence, people have a tendency to want to get on with a QFD project too fast; to try to determine the company measures before the customer requirements, for example.

It's also important that QFD be integrated into everyday business operations. Otherwise, it will just be an added task. In some cases, of course, QFD may have to be modified in order to make it fit; pure QFD isn't always practical.

At GE Aircraft Engines, for example, QFD is being used in conjunction with the engine development cycle (EDC), a highly structured approach to improved product development cycle times. GE Aircraft Engine's Kirkpatrick (1993) explained that QFD complements the EDC, which defines the best practices to follow when executing a design, by serving as a transitional tool that helps bring about necessary behavioral change.

GE Aircraft Engines has begun applying QFD to the CF6-8OC2, a 60,000-pound thrust-class turbofan engine that is used on almost all wide-body airplanes. According to Kirkpatrick, knowledge and improvements obtained from the CS6-8OC2 QFD project will be transferred to other GE Aircraft Engine designs as well, such as the new GE90, which will be used on the Boeing 777, and the company's military applications. In addition, a small QFD project that evolved from an earlier pilot is focusing on communication and data sharing between GE Aircraft Engines and its commercial customers.

Kirkpatrick stressed that he views QFD as a tool, not a process. "You do not have to use it everywhere—rather, use it where appropriate and adapt it. QFD is very adaptable, yet most people look at it as being very rigid." He also reported that QFD is now receiving high-level recognition at GE Aircraft Engines, but that it took nine months of hard work and a good amount of struggle to get there. This is not unusual. QFD requires patience from both project team members and upper management. It won't yield a quick return; but it will provide long-lasting benefits.

GM's Schaefer (1988) asked,

> Will the American engineer and the American manager have the patience to do QFD, or will the pressures of the everyday job win out? Up to this point, there's been a great deal of interest and curiosity. But when that wears off will we take the time to fill out the matrices? I'd like to think that we'll have the perseverance to do the job properly.

As Schaefer indicated, QFD—which GM's Saturn Corporation has successfully applied—also requires discipline. It brings discipline to an organization, but it also asks for discipline in return. It requires that people work together and pay strict attention to detail, which isn't always easy.

According to the CJQCA's Akashi Fukuhara (1988), who consults regularly with a variety of companies on QFD implementation, manufacturing must be more consistently drawn into the total product development process. "Most of the U.S. companies I've talked to have only involved quality assurance and design people—manufacturing and manufacturing engineering people are not yet involved in the QFD process," he explained.

"Getting the operations people to come to the party is a bit of a problem," agreed Schaefer. "Right now, QFD is more engineering driven. But manufacturing is beginning to participate. We've got some people within the manufacturing organization who are totally obsessed with it."

KEEPING THE MOMENTUM

The QFD pioneers initially interviewed for this book had their roots in the automotive industry, where QFD was first introduced. Several of the new QFD practitioners contacted for this second edition, however, were from different corporate sectors. Interest in QFD has spread to a wide array of industries and applications, and despite the challenges previously described, more QFD case studies are being completed by U.S. companies.

Although much of the information contained within these case studies is proprietary, questions and answers regarding QFD's mechanics, benefits, and troublesome aspects should be shared with others, as should case studies whenever possible. As ITT's Michael E. Chupa (1988) noted, promoting QFD without relying on success stories is a definite challenge, yet a company can't afford to divulge the results of a QFD case study to its competitors.

Public seminars, workshops, and conferences help address this problem, however, as do in-house seminars and symposia. QFD user groups are also extremely beneficial.

Companies with burgeoning workloads must work hard to keep QFD going strong. In such situations, upper management support is even more of a prerequisite to success. In fact, the level of management support necessary to move QFD through the organization is reciprocal to the size of the company. In small companies, upper management support is essential. In larger companies, however, QFD often results from a middle-up initiative; that is, rebel bands of QFD users who promote the process by show and tell. As GE Aircraft Engine's Kirkpatrick noted, individual applications of QFD can be effective advertisements for QFD when they prove to be successful.

Chrysler's Dika (1993) shared a similar sentiment.

> In most cases, if you wait for it to start at the top, you will be waiting for a long time. In practice, change has to start where there is passion and there are people willing to drive for change. . . . If QFD can be offered as a tool to help people do what their boss has just asked them to do, it is much easier to sell. The job of the facilitator is to be prepared to say the right thing, at the right time, with authority, and to be as clear as possible as to the outcome and the methods.

In both instances, top-down and middle-up, management astute enough to recognize the value of QFD should ensure that support staffs receive the training and hands-on experience necessary to make QFD a reality. Management that comes on strong for QFD—with no "if's, and's, or but's"—will see the greatest rewards. Management that sees no need or use for QFD should look again—within its own organizations and toward the future.

What else should you know before moving forward with QFD? Ten simple words: Find reasons to succeed with QFD, not excuses for failure.

Is Management Support Essential?

Top management should play an important role in the new product development process. By providing a broad, strategic direction for the company, top management helps define the process itself. This direction is the result of constant monitoring of the external environment, identifying competitive threats and market opportunities, and evaluation of company strengths and weaknesses.

Identifying competitive threats and market opportunities and evaluating company strengths and weaknesses are two of the things that QFD does best. To make QFD work, however, upper management needs to do two of the things it does best.

- Make a commitment to an innovative and rewarding activity.
- Delegate authority to the individuals best suited to make that activity happen.

QFD does require a definite time commitment, one that spans the entire product development cycle. By providing the QFD project team with the time to do QFD right, management sets the stage for a successful application.

Managing for success in QFD also entails asking the right questions. These include the following:

- How was the voice of the customer determined?
- How were the company measures determined? (Remember to challenge the usual in-house standards.)
- How do we compare with our competition?
- What opportunities can we identify to gain a competitive edge?
- What further information do we need and how can we acquire it?
- What trade-off decisions need to be made?
- What can I do to help?

There are a number of things management can do to help, most importantly, ensuring that the work environment is conducive to success. QFD is at its best in a progressive environment that fosters creative endeavors and information sharing. Although QFD will most likely be directed by a representative of design engineering, it will involve all product-oriented arms of the company.

Management should also push for success with QFD—but not too hard. QFD is essentially a learn-by-doing experience. With the proper encouragement, team members will push the limits of their own learning curve.

Management should provide QFD project team members—or, at the very least, the team leader—with professional instruction in QFD. This instruction will enhance QFD awareness and facilitation, as well as incite further interest in the process.

With a QFD champion in the ranks and on the management team, QFD can help your company improve its competitive ranking.

Tips from QFD Pioneers

If a company's healthy and getting new business, the best way to get started is by asking how QFD fits into the organizational scheme of things. The first exercise will be a real learning experience, but you won't be disappointed with the results.

—Michael E. Chupa, vice president of marketing,
ITT Automotive, Auburn Hills, Michigan

QFD will help us overcome departmentalization—is also one of the barriers that precludes its wide-scale implementation. The first step is to organize a multifunctional team. All functions must be represented.

—Dr. Don Clausing, Bernard Gordon Adjunct Professor
of Engineering Innovation and Practice,
Massachusetts Institute of Technology,
Cambridge, Massachusetts

We used a two-part strategy to get QFD started: awareness sessions and facilitation case studies. Each awareness session consisted of a one-hour pitch—20 minutes of prepared presentation and 40 minutes of questions and answers. We approached people at all levels—top executives, middle management, and line engineers.

—Robert J. Dika, quality planning executive—
small car platform, customer satisfaction
and vehicle quality office at Chrysler Corporation,
Auburn Hills, Michigan

You have to be patient—it's going to take some time. You can't expect to implement QFD across the board all at once. That's unrealistic—you just won't have the people or time to do that. Be selective in picking your QFD application and then take it step by step.

—James T. Gipprich, sales manager
of light duty vehicle brake group,
Kelsey–Hayes Company, Romulus, Michigan

You need short-term education of the technique followed by hands-on experience implementing it. You also need access to someone who can review what you've done and help correct and redirect your efforts. Your first QFD project will then be a meaningful exercise.

—Calvin W. Gray, president, Handy & Harmon Automotive Group, Auburn Hills, Michigan

Get an example of how QFD's being used so you have a role model in your company and then expand from there. Americans like to see things in actual practice—testimonies that say it's better than what we've been doing. We need good examples in the public domain.

—Walton M. Hancock, William Clay Ford Professor of Product Manufacturing, University of Michigan, Ann Arbor, Michigan

QFD is like driving a car—you can't learn all about it in a classroom. You have to get behind the wheel and actually engage the clutch. You also have to be flexible in your approach; you have to develop the charts to meet your needs.

—Norman E. Morrell, corporate manager of quality-product reliability, The Budd Company, Troy, Michigan

It's important to have a team leader, as well as an outside resource person. The former needs to be a high-potential person who's very credible in the organization and open-minded enough to see different points of view. The latter should be a knowledgeable facilitator who keeps things moving without directing the project and has no direct interest in the particular project.

—George R. Perry, vice president and general manager, Siemens Automotive, Newport News, Virginia

To quote Mr. Fukuhara, you've got to "pick up the pencil." You can talk and talk and talk about it, but until you actually start defining the voice of the customer and creating a chart, it's all talk. You don't have to do things perfectly—there's a lot of benefit from just making an attempt.

—Robert M. Schaefer, director of systems engineering methods, General Motors Corporation, Warren, Michigan

Putting the information down and keeping it up to date is a fairly labor-intensive procedure that someone needs to be responsible for. That someone may vary among companies, depending on how they are organized.

—Raymond P. Smock, retired manager of advanced quality concepts development and product assurance, Ford North American Automotive Operations, Dearborn, Michigan

The mechanics of QFD can be learned from a seminar. The actual information that you get from the marketplace, however, really makes QFD meaningful. Formulating a questionnaire, sending it off, analyzing the results—these are very important aspects of QFD.

—Peter J. Soltis, retired senior technical specialist, product engineering, Kelsey–Hayes Company, Romulus, Michigan

REFERENCES

Clausing, D.P. 1988. Personal interview. Interviewed by Nancy Ryan.

Chupa, M.E. 1988. Personal interview. Interviewed by Nancy Ryan.

Dika, R.J. 1993. *QFD Implementation at Chrysler: The First Seven Years.* Paper read at the Fifth Symposium on Quality Function Deployment, 21-22 June, at Novi, Michigan.

Dika. R.J. 1988. Personal interview. Interviewed by Nancy Ryan.

Fukuhara, A. 1988. Personal interview. Interviewed by Nancy Ryan.

Gipprich, J.T. 1988. Personal interview. Interviewed by Nancy Ryan.

Gray, C.W. 1988. Personal interview. Interviewed by Nancy Ryan.

Hancock, W.M. 1988. Personal interview. Interviewed by Nancy Ryan.

Kirkpatrick, R.A. 1993. Telephone interview. Interviewed by Nancy Ryan.

Schaefer, R.M. 1988. Personal interview. Interviewed by Nancy Ryan.

Smock, R.P. 1988. Personal interview. Interviewed by Nancy Ryan.

Soltis, P.J. 1988. Personal interview. Interviewed by Nancy Ryan.

Appendix

Taguchi Methods™

As noted in chapter 4, Quality Function Deployment (QFD) is often applied in conjunction with a variety of other processes and methodologies, including Taguchi Methods. These combine engineering and statistical methods that achieve rapid improvements in cost and quality by optimizing product design and manufacturing processes.

This is partly because the two processes are very complementary in nature. Whereas QFD identifies the relationships between inputs (*how* items) and outputs (*what* items) as well as any conflicting inputs that must be balanced, Taguchi Methods then define the nature of those relationships and optimize the conflicting inputs (see Figure A–1). Furthermore, the methodology can be used to desensitize the outputs of uncontrollable inputs, which reduces performance variation. This reduction in variability results in reduced cost and improved performance and quality.

Taguchi Methods, which were developed by Dr. Genichi Taguchi, are both a philosophy and a collection of tools used to carry forth that philosophy (see Figure A–2). This philosophy is summarized by the following statements.

1. We cannot reduce cost without affecting quality.
2. We can improve quality without increasing cost.
3. We can reduce cost by improving quality.

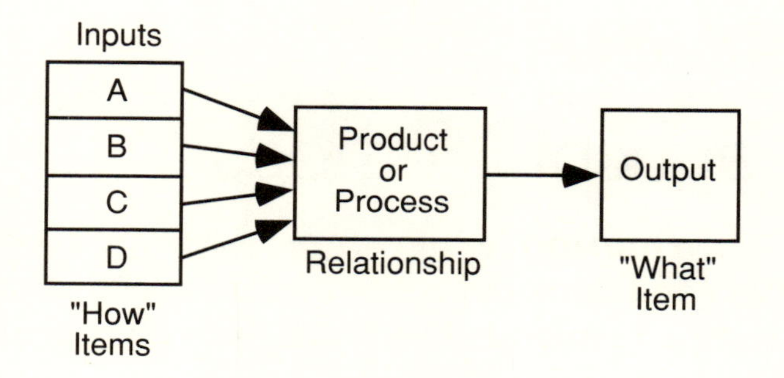

FIGURE A–1
QFD identifies the relationships of inputs and outputs, as well as conflicting inputs that would benefit from application of Taguchi Methods.

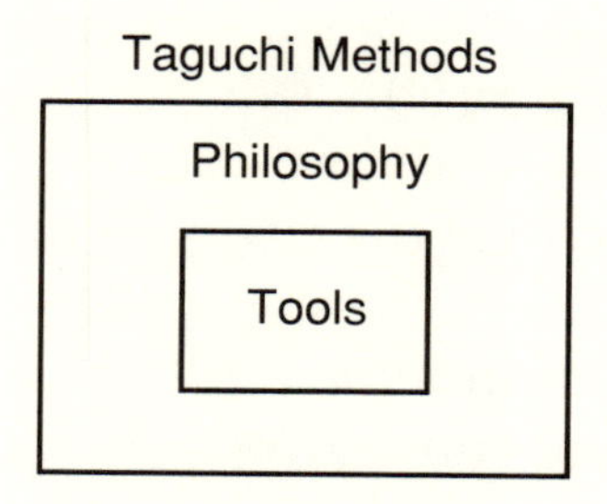

FIGURE A–2
Taguchi Methods are both a philosophy and a collection of tools used to carry forth that philosophy.

4. We can reduce cost by reducing variation. When we do so, performance and quality will automatically improve.

Taguchi has received four Deming Prizes for his work, as well as the Willard F. Rockwell Medal for Excellence in Technology. This award recognizes the generation, transfer, and application of technology for the betterment of humanity. It is from the International Technology Institute.

KEY TAGUCHI TERMS

The following key terms are pertinent to a clear understanding of Taguchi Methods.

Confirmation experiment An experiment run under the conditions defined as optimum by a previous experiment. It is intended to verify the experimental predictions.

Factor A parameter or variable that may impact product or process performance.

Linear graph A graphical representation of the assignment of factors to specific columns of an orthogonal array. Each linear graph is associated with one orthogonal array. A given array, however, can have several linear graphs.

Noise Any uncontrollable factor that causes product quality to vary. There are three kinds of noise: (1) external, or noise due to external causes, such as temperature, humidity, operator, vibration, and so on; (2) internal, or noise due to internal causes, such as wear, deterioration, and so on; and (3) product-to-product, or noise due to part-to-part variation.

Orthogonal array A matrix of numbers arranged in rows and columns. Each row represents the state of the factors in a given experiment. Each column represents a specific factor or condition that can be changed from test run to test run. The array is called orthogonal because the effects of the various factors in the experimental results can be separated from each other.

Parameter design The second of three design stages. During parameter design, nominal values of critical dimensions and characteristics are established to optimize performance at low cost.

Quality characteristics A characteristic of a product or process that defines the product/process quality; a measure of the degree of conformance to some known standard.

Quality Loss Function (QLF) A parabolic approximation of the quality loss that occurs when a quality characteristic deviates from its target value. The QLF is expressed in monetary units.

Robustness The condition of a product or process design that indicates that it functions with limited variability despite diverse and changing environmental conditions, wear, or component-to-component variation. A product or process is robust when it has limited or reduced functional variation in the presence of noise.

Signal-to-Noise (S/N) Ratio A quantity characterizing quality that originated in the communications field. When applied to design of experiments, the S/N Ratio is used to evaluate test equipment quality and project field quality performance from experimental data. The S/N Ratio is a measure that indicates how well variability has been minimized—the larger the ratio, the more robust the product is against noise.

System design The first of three design stages. During system design, scientific and engineering knowledge is applied to produce a functional prototype design.

This prototype is used to define the initial settings of product/process design characteristics.

Tolerance design The third of three design stages. Tolerance design is applied only if the design isn't acceptable at its optimum level following parameter design. During tolerance design, more costly materials or processes with tighter tolerances are considered.

WHEREIN LIES THE DIFFERENCE?

Taguchi disagrees with the conformance-to-specification-limits approach to quality. The difference between a product barely within specification limits (1 on Figure A–3) and a product barely out of specification limits (2 on Figure A–3) is small, yet one is considered good and the other bad. Rather, Taguchi Methods strive for minimal variation around target values without adding cost.

Taguchi defines quality as the characteristic that avoids "loss to society" from the time a product is shipped. When a product fails to function as expected, it imparts a loss to the customer. The customer may experience this loss on a small and/or large scale; for example, as the costs associated with repair or replacement of a product or as an environmental hazard created by a product.

When a product is overdesigned, there is a loss to the company. The product may even be inferior, because it may be heavier, less efficient, or larger than necessary. All loss is eventually experienced by the company through warranty costs, customer complaints, litigation, and eventual loss of reputation and market share.

According to Taguchi, quality is best when product characteristics are at target values (see Figure A–4). As

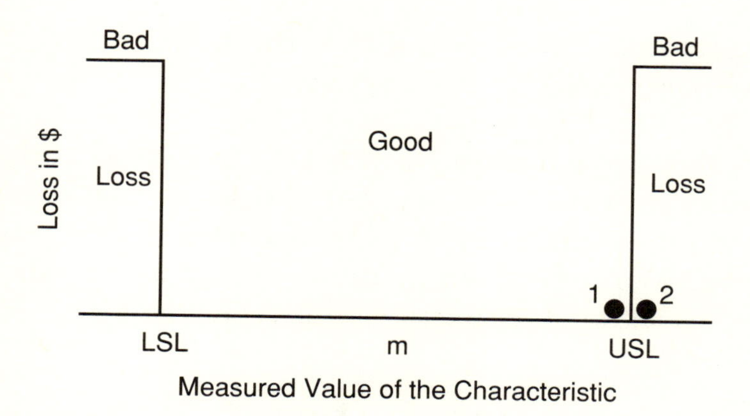

FIGURE A–3

The difference between a product barely within specification limits (1) and a product barely out of specification limits (2) is small, yet one is considered good and the other bad. According to Taguchi, minimal variation around the target value (*m*) results in reduced costs and consumer loss.

product characteristics deviate from target values, quality decreases and consumer dissatisfaction and loss increase.

The curve shown in Figure A-4 is known as the Quality Loss Function (QLF). The QLF is an enhanced cost-control system designed to quantitatively evaluate quality. It assesses quality loss due to deviation of a quality characteristic from its target value and then expresses this loss in monetary units.

The QLF quantifies annual cost savings as quality characteristics improve toward target values—even when within specification limits. It is an excellent tool for evaluating quality at the earliest stage of product/process development.

Thus, reducing sensitivity to variation is a main thrust of Taguchi Methods. This is accomplished by adjusting

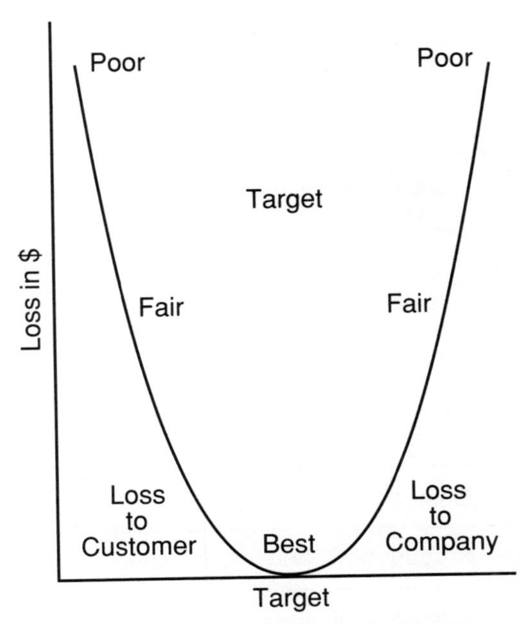

FIGURE A–4
As product characteristics deviate from target values, quality decreases and consumer dissatisfaction and loss increase.

factors that can be controlled in a way that minimizes the effects of factors that can't be controlled (see Figure A–5), which results in what Taguchi calls a robust design. Factors that can be controlled are called control factors; factors that are difficult, impossible, or expensive to control are called noise factors.

Taguchi has identified three types of noise factors, external, internal, and product-to-product, that generally cause product characteristics to deviate from target values. Thus, these cause variation and quality loss.

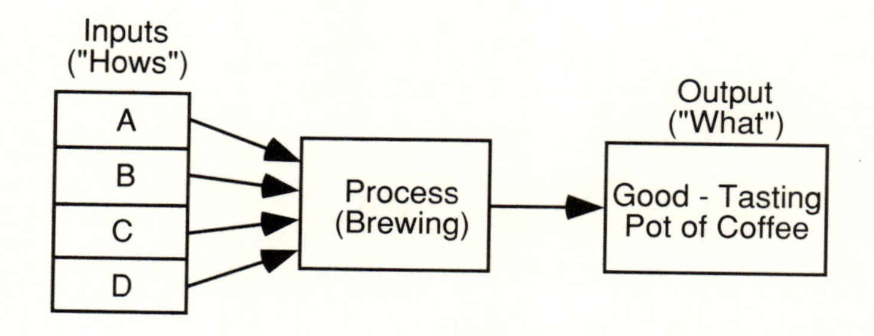

	Inputs ("Hows")	Choices 1	2	3
Control Factors	A — Choice of Coffee Bean	A_1	A_2	A_3
	B — Amount of Coffee	B_1	B_2	B_3
	C — Amount of Water	C_1	C_2	C_3
Noise Factor	D — Purity of Water	D_1	D_2	D_3

FIGURE A–5

In parameter design, the control factors (parameters) are adjusted to minimize the effect of noise factors. The goal is to find the combination of factors that gives the most stable and reliable performance at the lowest cost.

A SIMPLE EXAMPLE

Consider the following nontechnical analogy. There are a number of choices when it comes to the kind of coffee bean, amount of coffee, and amount of water used to make a pot of coffee. These can be thought of as control factors. Unless bottled or purified water is purchased, however, the purity of the water used to make the pot of coffee is hard to control. Hence, purity of water can be thought of as a noise factor.

In parameter design, the control factors are adjusted to minimize the effect of the noise factors. In this simple example, this means finding a cost-effective combination of coffee bean, amount of coffee, and amount of water that tastes best with a variety of water types. The end goal is to make a good-tasting pot of coffee at a low cost.

Control factors that minimize the effects of noise factors, and hence reduce variation, are selected during parameter design. This is the design stage following system design (see Figure A–6). The goal of parameter design is to find the combination of materials, processes, and specifications that gives the most stable and reliable performance at the lowest cost through experimentation.

This is accomplished by maximizing a measure called the Signal-to-Noise (S/N) Ratio. The S/N Ratio, which

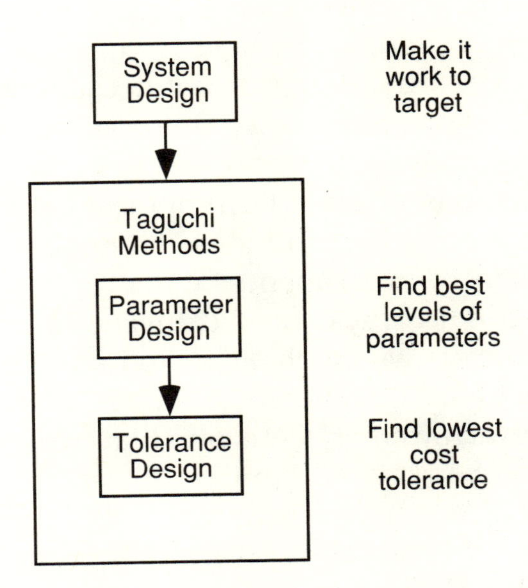

FIGURE A–6

Parameter design follows system design, the creation of a prototype design. If the design isn't acceptable following parameter design, tolerance design (which usually adds cost) is applied.

originated in the communications field (the signal representing the desired output; the noise, whatever gets in the signal's way), is an objective, statistical measure of performance and the effect of noise factors on performance. The S/N Ratio measures the stability of a quality characteristic's performance. The QLF is then used to evaluate the effect of that stability in monetary units. High performance, a large S/N Ratio, implies low loss, which will be measured by the QLF. The larger the ratio, the more robust the product will be against noise.

PEAK PERFORMANCE, LOW COST

When parameter design is complete, the design should be at its peak performance at the lowest cost. It is very important to run a confirmation experiment to verify that the experimental results can hold up in production and the marketplace. If the design is not acceptable following parameter design, more costly materials or processes with tighter tolerances can be applied to it.

Taguchi Methods utilize a form of experimental design that is much more suited to industrial applications than classical design of experiments. While Taguchi did not invent orthogonal arrays, he did invent linear graphs, which make the arrays much easier to use. Taguchi uses linear graphs to assign different numbers of test settings to factors. The arrays then focus on an average effect that occurs as other conditions change.

Orthogonal arrays have a pair-wise balancing property. The level for each factor occurs the same number of times as levels for all other factors (see Figure A–7). This minimizes the number of required experimental runs.

Taguchi Methods can be used for both off-line (product- and process-design optimization) and on-line quality

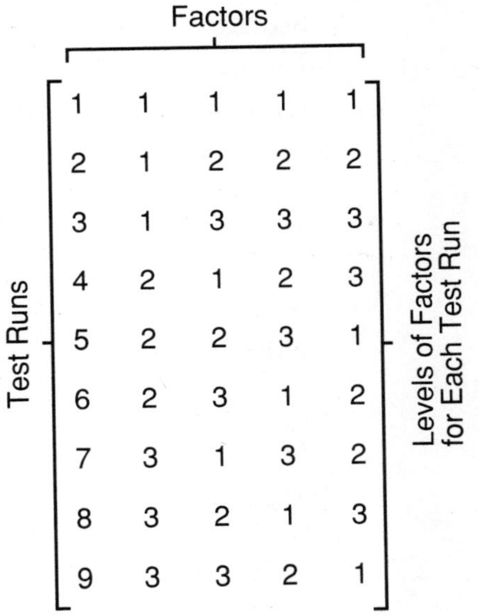

Factors

Test Runs

1	1	1	1	1
2	1	2	2	2
3	1	3	3	3
4	2	1	2	3
5	2	2	3	1
6	2	3	1	2
7	3	1	3	2
8	3	2	1	3
9	3	3	2	1

Levels of Factors for Each Test Run

	A Coffee Bean	B Amt. of Coffee	C Amt. of Water	D Undecided
1	1	1	1	1
2	1	2	2	2
3	1	3	3	3
4	2	1	2	3
5	2	2	3	1
6	2	3	1	2
7	3	1	3	2
8	3	2	1	3
9	3	3	2	1

FIGURE A–7

This orthogonal array, the L9, features nine test runs with four different factors, each of which is run at three different settings. Each factor is exposed to each setting an equal number of times, which ensures reproducibility of results. Control factors for the pot-of-coffee example are shown at right. (The noise factor would be placed in an adjacent outer array.)

control activities. The majority of Taguchi Methods applications in the United States involve off-line quality control. The challenges are to increase the use of Taguchi Methods for on-line quality control and to use them in conjunction with QFD.

Glossary

Company measures Global measures, not specific to design, that translate the voice of the customer into broad, actionable issues the company can work on in response to customer wants. Also called technical requirements.

Companywide Quality Control See Total Quality Management.

Competitive Assessment A benchmarking process for evaluating customer requirements and company measures that determines the performance of the competition.

Concurrent Engineering See simultaneous engineering. (The term *simultaneous engineering* is favored by the automotive industry, concurrent engineering by the defense industry.)

Correlation The corelationship between any two company measures. It represents a team judgment about whether action to improve one company measure will harm or help the other company measure.

Customer rating Customer opinion of the importance of each customer want, typically rated on a numerical scale.

Customer requirements Written representations of the voice of the customer; wants and needs of customers determined during interviews and through surveys.

Deployment The process of transforming information through the product development process into a saleable product.

Design for Assembly (DFA) A method of design simplification that focuses on reducing assembly complexity without sacrificing product performance. DFA was introduced by Geoffrey Boothroyd and Peter Dewhurst of the University of Massachusetts in 1980.

Design of Experiments (DOE) Statistical techniques used to discover the effects of factors upon desired results. Taguchi Methods use a simple subset of DOE to get engineering results quickly and without extensive statistical expertise.

Failure Mode and Effect Analysis (FMEA) A method for failure reduction that identifies potential product or process weakness, assesses risks, and determines preventive actions.

Fault Tree Analysis (FTA) A method of tracing failures through a chain of cause-and-effect relationships to their root cause. FTA can provide a means to estimate failure probability.

House of Quality A term used to describe the QFD product planning matrix designed to increase customer satisfaction to improve quality. The correlation matrix at the top gives the appearance of a house. Its purpose and appearance account for the term.

Matrix A rectangular chart where two sets of items are compared. One set is entered horizontally and the second set vertically.

Policy Management A process for developing achievable business plans and then deploying them throughout the company that closely resembles QFD. Also called Policy Deployment.

Pugh Concept Selection A creative method for comparing design concepts that seeks to synthesize a single concept better than any of the initial concepts. Devised by Dr. Stuart Pugh, a British engineer.

Relationships Symbolic representations of cause-effect phenomenon.

Reverse Fault Tree Analysis (RFTA) A mirror image of Fault Tree Analysis that focuses on drivers for success rather than causes of failure. An FTA can be logically transformed into an RFTA.

Robust design The use of Taguchi Methods to define and optimize product function.

Seven Basic Tools A set of seven tools primarily used to derive information from quantitative data. The tools are cause-and-effect diagram, Pareto analysis or diagram, check sheet, scatter diagram, histogram, stratification, and control chart.

Seven Management Tools A set of seven tools primarily used to derive information from qualitative data. The tools are affinity diagram, relationship digraph, tree diagram, matrix chart, matrix data analysis, process design program chart, and arrow diagram.

Simultaneous engineering A way of simultaneously designing products, and the processes for manufacturing those products, through the use of parallel efforts to ensure manufacturability and reduce cycle time. Also called Concurrent Engineering.

Taguchi Methods™ An engineering approach that optimizes the product to achieve stable, cost-effective performance.

Target value A goal established for each company measure. It represents the team judgment of the level of achievement that must be attained for the company measure to satisfy customers.

Technical requirements See company measures.

Total Quality Management (TQM) A management approach that integrates fundamental management techniques, existing improvement efforts, and technical tools to transform a company into a continuously improving organization. Also called Companywide Quality Control and Total Quality Control.

Value Analysis/Value Engineering (VAVE) A method of design simplification that focuses on reducing cost while maintaining or improving product functionality. This was introduced by Larry Miles of General Electric in 1947.

Voice of the customer The wants and needs that customers express during interviews and surveys relative to products and service performance.

Bibliography

Akao, Y. ed. 1990. *Quality Function Deployment: Integrating Customer Requirements into Product Design.* Cambridge, Mass.: Productivity Press.

Bossert, J.L. 1990. *QFD: A Practitioner's Approach.* Milwaukee: ASQC Quality Press.

Camp: R.C. 1989. *Benchmarking: The Search for Industry Best Practices that Lead to Superior Performance.* Milwaukee: ASQC Quality Press and White Plains, N.Y.: Quality Resources.

Clausing, D.P., and S. Pugh. 1991. "Enhanced Quality Function Deployment." Paper presented at Design and Productivity International Conference.

Cohen, L. 1988. Quality Function Deployment: An Application Perspective from Digital Equipment Corporation. *National Productivity Review* (Summer).

Covey, S.R. 1989. *The Seven Habits of Highly Effective People: Powerful Lessons in Personal Change.* New York: Simon & Schuster.

Hauser, J.R., and D.P. Clausing. 1988. The House of Quality. *Harvard Business Review* (May–June):63–73.

Hofmeister, K.R. 1990. QFD in the Service Environment. *The ASI Journal* 3, no. 2.

King, B. 1987. *Better Designs in Half the Time: Implementing QFD in America.* Methuen, Mass.: GOAL/QPC.

LeBouef, M. 1989. *How To Win Customers and Keep Them for Life.* New York: Berkeley Publishing Group.

Marsh, S., J.W. Moran, S. Nakul, and G. Hoffherr. 1991. *Facilitating and Training in Quality Function Deployment.* Methuen: Mass.: GOAL/QPC.

Phadke, M. 1989. *Quality Engineering and Robust Design.* Englewood Cliffs, N.J.: Prentice-Hall.

Pugh, S. 1981. Concept Selection—The Method That Works. In Proceedings of ICED.

———. 1990. *Total Design: Integrated Methods for Successful Product Engineering.* Reading, Mass.: Addison-Wesley.

Quality Function Deployment and the Competitive Challenge: A Special Nine-Part Video Series with Bill Eureka. 1989. Dearborn, Mich.: ASI Press. Videocassette.

Quality Function Deployment: The Budd Company Case Study. 1987. Dearborn, Mich.: ASI Press. Videocassette.

Quality Function Deployment Executive Briefing. 1992. Dearborn, Mich.: ASI Press.

Quality Function Deployment for Continuous Batch Process Implementation Manual. 1992. Dearborn, Mich.: ASI Press.

Quality Function Deployment for Hardware Implementation Manual. 1992. Dearborn, Mich.: ASI Press.

Quality Function Deployment for Service Implementation Manual. 1992. Dearborn, Mich.: ASI Press.

Quality Function Deployment: The Kelsey–Hayes Case Study. 1987. Dearborn, Mich.: ASI Press. Videocassette.

Rosenthal, S.R. 1992. *Effective Product Design and Development: How To Cut Lead Time and Increase Customer Satisfaction*. Homewood, Ill.: BusinessOne Irwin.

Ryan, N. ed. 1988. *Taguchi Methods and QFD: Hows and Whys for Management*. Dearborn, Mich.: ASI Press.

Scholtes, P.R. 1988. *The Team Handbook: How To Use Teams To Improve Quality*. Madison, Wis.: Joiner Associates.

Shunk, D.L. 1992. *Integrated Process Design and Development*. Homewood, Ill.: BusinessOne Irwin.

Sudman, S., and N.E. Bradburn. 1982. *Asking Questions: A Practical Guide to Questionnaire Design*. San Francisco: Jossey-Bass.

Sullivan, L.P. 1986. Quality Function Deployment. *Quality Progress* (June): 39–50.

ter Haar, S., D.P. Clausing, and S.D. Eppinger. 1993. Integration of Quality Function Deployment and the Design Structure Matrix. Cambridge, Mass.: Massachusetts Institute of Technology.

Transations, First through Fifth Symposiums on Quality Function Deployment. 1989–1993. Dearborn, Mich.: ASI Press and Methuen, Mass.: GOAL/QPC.

Wheelwright, S.C., and K.B. Clark. 1992. *Revolutionizing Product Development: Quantum Leaps in Speed, Efficiency, and Quality*. New York: Free Press.

Ziaja, H.J. 1990. Total Product Quality Model. Paper read at ASQC 44th Annual Quality Congress.

Index

Other books of interest to you from Irwin Professional Publishing . . .

THE PROCESS–DRIVEN BUSINESS

Managerial Perspectives on Policy Management

William E. Eureka and Nancy E. Ryan
U.S. Distribution for American Supplier Institute Press

Examines the many benefits of Policy Management—a progressive business planning technique that uses process and systems evaluations to generate long-lasting improvements. Includes tips for getting started with Policy Management and ensuring its effectiveness. (138 pages)
ISBN: 0-941243-12-5

QUALITY BY DESIGN, SECOND EDITION

Taguchi Methods and U.S. Industry

Lance A. Ealey
Co-published with American Supplier Institute Press

Now's there's a simple way for any manager to understand Taguchi Methods! *Quality by Design* provides a nontechnical explanation, showing how Taguchi Methods can lead to rapid improvements in costs and quality. Includes a one-on-one interview with Dr. Taguchi! (330 pages)
ISBN: 1-55623-970-X

GLOBAL QUALITY

A Synthesis of the World's Best Management Methods

Richard Tabor Greene
Co-published with ASQC Quality Press

Finally, a book that organizes the chaos of quality improvement techniques—so you can select the most appropriate system for your organization. Inside, you'll find the 24 quality approaches used worldwide, the essentials of process reengineering, software (groupware) techniques, and much more.

Also reveals what the top 1 percent of quality professionals are doing to advance quality, including seven new quality improvement techniques being tested in Japan! (784 pages)
ISBN: 1-55623-915-7

MANAGEMENT OF QUALITY

Strategies to Improve Quality and the Bottom Line

Jack Hagan
Co-published with ASQC Quality Press
Presents core concepts of quality improvement in a language any manager can understand: the language of business. Jack Hagan examines the true state of quality today and shows you how to link its basic principles to your company's business strategy—including its profit-oriented objectives. (163 pages)
ISBN: 1-55623-924-6

Available at fine bookstores and libraries everywhere.